Lecture Notes in Mathematics

A collection of informal reports and seminars
Edited by A. Dold, Heidelberg and B. Eckmann, Zürich

324

Kurt Kreith
University of California, Davis, CA/USA

Oscillation Theory

Springer-Verlag
Berlin · Heidelberg · New York 1973

AMS Subject Classifications (1970): 34B25, 34C10, 34G05, 35B05, 35O15, 35L10

ISBN 3-540-06258-0 Springer-Verlag Berlin · Heidelberg · New York
ISBN 0-387-06258-0 Springer-Verlag New York · Heidelberg · Berlin

Offsetdruck: Julius Beltz, Hemsbach/Bergstr.

PREFACE

These notes were written in conjunction with a series of lectures given at Chelsea College of the University of London in Fall, 1972.

The author is indebted to Dr. M. S. P. Eastham for the invitation to spend a sabbatical year at Chelsea College and to the Science Research Council for its generous support. Special thanks are also due to Mrs. Ida Orahood for her expert typing and help in the preparation of the manuscript and to Professor Curtis Travis for a number of helpful suggestions.

CONTENTS

CHAPTER 0

INTRODUCTION

If $u(x)$ and $v(x)$ are nontrivial solutions of

$$-u'' + pu = 0 \tag{0.1}$$

$$-v'' + Pv = 0 \tag{0.2}$$

for constant p and P, then

(i) $P < p \Rightarrow$ successive zeros of $u(x)$ are separated by a zero of $v(x)$.

(ii) $p \geq 0 \Rightarrow$ $u(x)$ has at most one zero on any interval $[x_o,\infty)$.

(iii) $P < 0 \Rightarrow$ $v(x)$ has a zero on every interval $[x_o,\infty)$.

These observations constitute the simplest examples of separation, nonoscillation, and oscillation theorems, respectively. They are also the motivation for a large and growing body of mathematical literature.

The purpose of these notes is to present some recent developments in this theory. This survey is by no means complete; in fact a conscious effort has been made to minimize the overlap with other recent surveys in this area by Barrett [1], Coppel [1], Swanson [1], and Willett [1]. Thus, while Sturmian Theorems for elliptic partial differential equations are dealt with by Swanson [1], our emphasis in Chapter 3 is on material that has appeared since the publication of Swanson's book. Chapter 4 is an attempt to establish relations between oscillation theory and other topics in analysis. A variety of recent abstract formulations of oscillation theory is developed in Chapter 5, but no attempt is made to deal with the general theory of Hamiltonian systems which is covered by Coppel [1] and Reid [1], [2]. Much of the complex oscillation theory of Chapter 6 is covered by Ince [1] and Hille [1], but it is included here to clarify some of the difficulties encountered in the part of Chapter 7 which deals with nonselfadjoint differential equations. Chapter 8 deals with some special results for fourth order equations which do not have obvious generalizations to the more general case of equations of even order considered in Chapter 7.

The material presented can be approached in a variety of ways. Some authors have made extensive use of Riccati equations, variational techniques, or differential inequalities in obtaining related results. The principal virtue of the techniques presented here is that they tend to be rather direct generalizations of the techniques used in the classical development of the theory. The notes represent a particular point of view of the material covered, and the interested reader is encouraged to consult the references for alternate techniques which have been successfully employed.

CHAPTER 1

THE CLASSICAL THEORY

1. Sturm's Comparison Theorem

Most of the classical results in oscillation theory are formulated for solutions of selfadjoint Sturm-Liouville equations of the form

$$(1.1) \qquad -(p_1(x)u')' + p_o(x)u = 0$$

$$(1.2) \qquad -(P_1(x)v')' + P_o(x)v = 0$$

where the $p_k(x)$ and $P_k(x)$ are of class C^k and $p_1(x)$ and $P_1(x)$ are positive on an appropriate interval $\mathcal{J}$. The starting point for this theory is the well known comparison theorem of C. Sturm [1] discovered in 1836.

1.1 Theorem. *If* α, β *are consecutive zeros of a nontrivial solution* $u(x)$ *of* (1.1) *and if*

(i) $\qquad P_1(x) \equiv p_1(x)$

(ii) $\qquad P_o(x) \le p_o(x) \quad ; \quad P_o(x) \not\equiv p_o(x)$

for $x \in \mathcal{J} = [\alpha,\beta]$, *then every solution* $v(x)$ *of* (1.2) *has a zero in* (α,β).

Proof. If $u(x)$ and $v(x)$ are solutions of (1.1) and (1.2), respectively, then

$$(1.3) \qquad \frac{d}{dx}[vp_1u'-uP_1v'] = (p_o-P_o)uv.$$

If $u(\alpha) = u(\beta) = 0$, then an integration yields

$$vp_1u'\Big|_{x=\alpha}^{x=\beta} = \int_\alpha^\beta (p_o-P_o)uv\, dx.$$

Assuming $u(x) > 0$ on (α,β), $u'(\alpha) > 0$, and $u'(\beta) < 0$, it follows from (1.3) that $v(x)$ cannot be of constant sign on (α,β). An analogous argument is valid if $u(x) < 0$ on (α,β).

2. Physical Interpretations

This theorem has some simple physical interpretations which will motivate later generalizations.

(a) Vibrating String. Suppose $P_1(x) \equiv p_1(x) \equiv 1$ and

$$-p_o(x) = \delta_1(x) > 0$$

$$-P_o(x) = \delta_2(x) > 0.$$

If $u(\alpha) = u(\beta) = 0$, then the eigenvalue problem

$$u'' + \lambda\delta_1(x)u = 0$$

$$u(\alpha) = u(\beta) = 0$$

has eigenvalue $\lambda_1 = 1$, corresponding to a string of density $\delta_1(x)$ with fundamental frequency of unity. A second string with density $\delta_2(x) \geq \delta_1(x)$, $\delta_2(x) \not\equiv \delta_1(x)$ will have a lower fundamental frequency. Any solution of $v'' + \delta_2(x)v = 0$ can be thought of as an eigenfunction of

$$v'' + \mu v = 0$$

$$v'(\alpha) - \tau_1 v(\alpha) = 0$$

$$v'(\beta) + \tau_2 v(\beta) = 0$$

corresponding to $\mu_k = 1$. But by the physical argument above, the eigenvalue $\mu_k = 1$ will not be the first eigenvalue. Since the first eigenfunction is the only one of constant sign, it follows that $v(x)$ changes sign in (α,β). This argument applied to membranes suggests that a Sturmian comparison should also be valid for elliptic equations.

(b) Simple Harmonic Motion. Let $P_1(x) \equiv p_1(x) \equiv 1$ and x denote time. Given a particle with one degree freedom which is attracted to $u = 0$ with a force $f(x) = -p_o(x)u$, the motion of this particle is described by $u'' + f(x)u = 0$. Another particle attracted to $v = 0$ with a force $F(x) = -P_o(x)v$ satisfying $F(x) \geq f(x)$,

$F(x) \not\equiv f(x)$, will oscillate more rapidly. That is, if $u(\alpha) = u(\beta) = 0$, then during the interval (α,β) the second particle will pass through $v = 0$ at least once. Applying similar reasoning to two strings oscillating about an equilibrium line should yield Sturmian theorems for hyperbolic equations.

(c) Planetary Motion. The simple harmonic motion argument does not hold if there is more than one degree of freedom. Thus if a particle is free to move in the plane under a central force field, the strength of attraction to the origin will not determine the frequency with which the particle passes through the origin. This observation constitutes a physical interpretation of Sturmian nonoscillation theory for complex valued solutions of (1.1) and (1.2) and will be discussed further in Chapter 6.

3. The Sturm-Picone Theorem

Elaboration of the physical interpretations (a) and (b) suggests that the limitation $P_1(x) \equiv p_1(x)$ is not essential to the theory. This restriction was removed in 1909 by M. Picone [1] who made use of the identity

$$\frac{d}{dx}\left[\frac{u}{v}(vp_1u'-uP_1v')\right] = u(p_1u')' - \frac{u^2}{v}(P_1v')' + (p_1-P_1)u'^2 + P_1\left(u' - \frac{u}{v}v'\right)^2, \tag{1.4}$$

which is valid whenever u, v, p_1u', P_1v' are differentiable and $v(x) \neq 0$. This identity led Picone to the following generalization of Sturm's theorem.

1.2 Theorem. *If α, β are consecutive zeros of a nontrivial solution $u(x)$ of (1.1) and if*

(i) $$0 < P_1(x) \leq p_1(x)$$

(ii) $$P_o(x) \leq p_o(x)$$

for $x \in [\alpha,\beta]$, then every solution $v(x)$ of (1.2) has a zero in $[\alpha,\beta]$.

Proof. If $u(x)$ and $v(x)$ are solutions of (1.1) and (1.2), respectively, and $v(x) \neq 0$ for $x \in [\alpha,\beta]$, then (1.4) yields

$$\frac{d}{dx}\left[\frac{u}{v}(vp_1u'-uP_1v')\right] = (p_o-P_o)u^2 + (p_1-P_1)u'^2$$

$$+ P_1\left(u' - \frac{u}{v}v'\right)^2.$$

Integrating from α to β yields

$$0 \geq \int_\alpha^\beta \left[(p_o-P_o)u^2 + (p_1-P_1)u'^2\right] dx$$

which is a contradiction unless $p_o(x) \equiv P_o(x)$, $p_1(x) \equiv P_1(x)$, and $u'(x) - \frac{u(x)}{v(x)}v'(x) \equiv 0$. The last identity implies that $\frac{u(x)}{v(x)} \equiv$ constant, so that $v(\alpha) = v(\beta) = 0$.

Remarks.

1. The conclusion $v(x)$ has a zero in $[\alpha,\beta]$ can be strengthened to $v(x)$ has a zero in $[\alpha,\beta)$ and in $(\alpha,\beta]$ by considering $\lim_{x \downarrow \alpha} \frac{u}{v}$ and $\lim_{x \uparrow \beta} \frac{u}{v}$ and noting that (1.4) remains valid as long as $v(x) \neq 0$ on (α,β).

2. The assumption $u(\alpha) = u(\beta) = 0$ may be replaced by the more general conditions

$$u'(\alpha) - \sigma_1 u(\alpha) = 0$$

$$u'(\beta) + \sigma_2 u(\beta) = 0$$

where $\sigma_1 = +\infty$ denotes $u(\alpha) = 0$ and $\sigma_2 = +\infty$ denotes $u(\beta) = 0$. Theorem 1.2 then applies to solutions $v(x)$ of (1.2) satisfying

$$v'(\alpha) - \tau_1 v(\alpha) = 0$$

$$v'(\beta) + \tau_2 v(\beta) = 0$$

with $\tau_1 \geq \sigma_1$, $\tau_2 \geq \sigma_2$, and the additional assumption that $u(x)$ and $v(x)$ be linearly independent.

3. In the special case $P_1(x) \equiv p_1(x)$, $P_o(x) \equiv p_o(x)$, Theorem 1.2 becomes the Sturm separation theorem which asserts that the zeros of linearly independent

solutions of (1.1) are interlaced.

4. Oscillation Criteria

With the comparison theorems of Sturm and Picone as tools, one can establish a number of oscillation and nonoscillation criteria. An equation of the form (1.1) is said to be oscillatory at ∞ if every (some) nontrivial solution has a zero in every interval of the form $[\alpha,\infty)$. It is nonoscillatory at ∞ if some (every) nontrivial solution has a finite number of zeros in an interval of the form $[\alpha,\infty)$. It is said to be disconjugate on an interval $\mathcal{J}$ if no solution has more than one zero in $\mathcal{J}$. By comparing an equation of the form (1.2) with an equation of the form (1.1) which is known to be oscillatory, one obtains oscillation criteria for (1.2). Nonoscillation and disconjugacy criteria can be established for (1.1) in an analogous way. Of special interest here are the Euler equations which can be used for comparison in the application of this technique.

While the use of comparison theorems leads to a number of useful oscillation and nonoscillation criteria, such as those of Kneser [1], the most interesting and sensitive tests cannot be obtained in this way. The difficulty is that the pointwise inequalities required of the coefficients in Theorems 1.1 and 1.2 do not yield good results when the coefficients themselves are oscillatory. To deal with this situation one seeks criteria for comparison, oscillation, and nonoscillation theorems which involve inequalities among certain definite or indefinite integrals of the coefficients. While we shall not seek to survey the extensive literature on this subject in depth (see Swanson [1] and Willett [1] for more details) we shall present some important results of this type.

By way of motivation, we note that the special equation

$$\left(\frac{1}{p(x)}u'\right)' + p(x)u = 0 \tag{1.5}$$

has general solution

$$u(x) = A \sin\left[\int_\alpha^x p(\xi)\, d\xi + B\right].$$

Thus (1.5) is oscillatory at ∞ if $\int_\alpha^\infty p(x)\,dx = +\infty$ and nonoscillatory if $\int_\alpha^\infty p(x)\,dx < \infty$. An important related result for (1.1) is due to Leighton [1].

1.3 Theorem. *If* $\int_\alpha^\infty \frac{1}{p_1(x)}\,dx = +\infty$ *and* $\int_\alpha^\infty p_o(x)\,dx = -\infty$, *then* (1.1) *is oscillatory at* ∞.

Proof. The proof depends on an important relation between the Sturm-Liouville equation (1.1) and a special Riccati equation. Defining $h(x) = -p_1u'/u$ for a solution $u(x)$ of (1.1), one obtains

$$h' = -p_o + \frac{1}{p_1}h^2. \tag{1.6}$$

This transformation relates the zeros of $u(x)$ to the singularities of a solution of (1.6) and constitutes a basic tool in oscillation theory. Suppose to the contrary that (1.1) is nonoscillatory. If $u(x) > 0$ on $[\beta,\infty)$ then $h(x)$ defined as above is a solution of (1.6) on (β,∞). Since $\int_\alpha^\infty p_o(x)\,dx = -\infty$, we can find β and γ such that $\alpha < \beta < \gamma < \infty$ and $h(\beta) - \int_\beta^x p_o(\xi)\,d\xi > 0$ on $[\gamma,\infty)$. Defining $g(x) = \int_\beta^x \frac{1}{p_1(\xi)}h^2(\xi)\,d\xi$, we have $h(x) > g(x)$ on $[\gamma,\infty)$. From the definition of $g(x)$, $g'(x) > \frac{1}{p_1(x)}g^2(x)$ on $[\gamma,\infty)$ so that

$$\int_\gamma^\infty \frac{1}{p_1(x)}\,dx < \frac{1}{g(\gamma)} < \infty,$$

and this completes the proof.

By way of partial converse to Theorem 1.3, we mention the following theorem whose proof will be given in the next section.

1.4 Theorem. *If* $\int_\alpha^\infty \frac{1}{p_1(x)}\,dx < \infty$ *and* $\int_\alpha^\infty |p_o(x)|\,dx < \infty$, *then* (1.1) *is nonoscillatory at* ∞.

Related results can be found in Swanson [1] and elsewhere.

5. The Prüfer Transformation

Another classical approach to oscillation theory which is motivated by the special equation (1.5) is the transformation of Prüfer [1]. Given a solution $u(x)$ of (1.1), we seek to represent this solution in the form $u(x) = r(x) \sin \theta(x)$, where $r(x)$ and $\theta(x)$ are to be chosen appropriately. The existence of such $r(x)$ and $\theta(x)$ is not in doubt, but the question is whether they can be determined by means of equations which are useful in describing the behavior of $u(x)$. To that end we impose the additional requirement $p_1(x)u'(x) = r(x) \cos \theta(x)$ and find that $\theta(x)$ and $r(x)$ must satisfy

$$\theta' = \frac{1}{p_1(x)} \cos^2 \theta - p_o(x) \sin^2 \theta \tag{1.7}$$

$$r' = r\left(\frac{1}{p_1} + p_o\right) \sin \theta \cos \theta. \tag{1.8}$$

In case $\frac{1}{p_1(x)} \equiv -p_o(x) = p(x)$, we get $\theta(x) = \int_\alpha^x p(\xi)\, d\xi$ as in (1.5). Since $u(x_o) = 0$ if and only if $\theta(x_o) \equiv 0 \pmod{\pi}$, (1.1) is oscillatory at ∞ if $\lim_{x\to\infty} \theta(x) = \infty$ while it is nonoscillatory at ∞ if $\lim_{x\to\infty} \theta(x)$ exists. It is possible to give an elementary proof of Theorem 1.2 using this transformation and properties of differential inequalities (see Birkhoff and Rota [1]). Further generalizations of this transformation will be considered in succeeding chapters.

As one application of the Prüfer transformation we sketch a proof of Theorem 1.4. Since $p_1(x) > 0$ on $\mathcal{I}$, $\theta'(x_o) > 0$ at every zero of $u(x)$. Therefore (1.1) is oscillatory at ∞ if and only if $\lim_{x\to\infty} \theta(x) = +\infty$. If the hypotheses of Theorem 1.4 are satisfied, then

$$\varphi' = \frac{1}{p_1(x)} + |p_o(x)| \quad ; \quad \varphi(\alpha) = \theta_o$$

has a bounded solution on $[\alpha,\infty)$ for every θ_o. Since

$$\left|\frac{1}{p_1(x)} \cos^2 \theta - p_o(x) \sin^2 \theta\right| \leq \frac{1}{p_1(x)} + |p_o(x)|$$

it follows that every solution of (1.7) is bounded and that (1.1) is nonoscillatory at ∞.

CHAPTER 2

NONSELFADJOINT EQUATIONS

1. A Generalized Sturm-Picone Theorem

There are several ways of transforming the nonselfadjoint equation $a(x)u'' + b(x)u' + c(x)u = 0$ into a selfadjoint form and establishing oscillation properties of such equations by using the techniques of Chapter 1. However these transformations do not apply to the more general equations to be considered later, and for this reason it is of interest to establish oscillation properties by applying the techniques of Chapter 1 directly to equations in nonselfadjoint form. We seek to do this first for the equations

$$-(p_1(x)u')' + q_o(x)u' + p_o(x)u = 0, \tag{2.1}$$

$$-(P_1(x)v')' + Q_o(x)v' + P_o(x)v = 0, \tag{2.2}$$

where the $p_k(x)$, $q_k(x)$, $P_k(x)$, $Q_k(x)$ are of class C^k and $p_1(x)$, $P_1(x)$ are positive on an appropriate interval $\mathcal{I}$.

Of primary interest will be a generalization of the Sturm-Picone theorem to these equations. Starting with the Picone identity (1.4) and making use of (2.1) and (2.2) we get

$$\begin{aligned} \frac{d}{dx}\left[\frac{u}{v}(vp_1u'-uP_1v')\right] &= (p_o-P_o)u^2 + q_ouu' \\ &\quad - Q_o\frac{u^2}{v}v' + (p_1-P_1)u'^2 + P_1(u' - \frac{u}{v}v')^2 \end{aligned} \tag{2.3}$$

whenever $v(x) \neq 0$. To assure the positivity of the right hand side we seek to "complete the square" by means of functions $g_1(x)$ and $g_2(x)$ for which

$$(p_1-P_1)u'^2 + (q_o-Q_o)uu' + g_1u^2 \geq 0$$

$$P_1(u' - \frac{u}{v}v')^2 + Q_ou(u' - \frac{u}{v}v') + g_2u^2 \geq 0.$$

These inequalities are satisfied if

$$g_1 = \frac{(q_o-Q_o)^2}{4(p_1-P_1)} \quad ; \quad g_2 = \frac{Q_o^2}{4P_1} \quad ,$$

and for these choices of g_1 and g_2 we have

$$\frac{d}{dx}\left[\frac{u}{v}(vp_1u'-uP_1v')\right] \geq \left[p_o - P_o - \frac{(q_o-Q_o)^2}{4(p_1-P_1)} - \frac{Q_o^2}{4P_1}\right]u^2.$$

This last inequality readily yields the following generalizations of Theorem 1.2.

2.1 Theorem. *If* α, β *are consecutive zeros of a nontrivial solution* $u(x)$ *of* (2.1) *and if*

(i) $\qquad 0 < P_1(x) \leq p_1(x)$

(ii) $\qquad 0 < P_1(x) < p_1(x)$ *whenever* $q_o(x) \neq Q_o(x)$

(iii) $\qquad P_o(x) + \dfrac{(q_o(x)-Q_o(x))^2}{4(p_1(x)-P_1(x))} + \dfrac{Q_o^2(x)}{4P_1(x)} \leq p_o(x)$

for $x \in [\alpha,\beta]$, *then every solution* $v(x)$ *of* (2.2) *has a zero in* $[\alpha,\beta]$.

2. A Device of Picard

The remarks following Theorem 2.1 remain valid in the present setting. In addition, it is of interest to note a variation of (2.3) which is based on a device due to Picard [1] and leads to different versions of Theorem 2.1. For any differentiable $f(x)$ we have

$$\frac{d}{dx}(u^2f) = 2uu'f + u^2f'.$$

Adding this to (2.3) yields

$$\frac{d}{dx}\left[\frac{u}{v}(vp_1u'-uP_1v') + u^2f\right] = (p_o-P_o+f')u^2 + (q_o-Q_o+2f)uu'$$

$$+ Q_ou(u' - \frac{u}{v}v') + (p_1-P_1)u'^2 + P_1(u' - \frac{u}{v}v')^2$$

where one can now seek to complete the square for various choices of f. Assuming q_o and Q_o are of class C^1, the choice $f = \frac{Q_o - q_o}{2}$ yields

$$\frac{d}{dx}\left[\frac{u}{v}(vp_1u' - uP_1v') + u^2\,\frac{Q_o - q_o}{2}\right] \geq \left[p_o - P_o - \frac{q_o' - Q_o'}{2} - \frac{Q_o^2}{4P_1}\right]u^2 + (p_1 - P_1)u'^2$$

and the following result.

2.2 Theorem. *If* α, β *are consecutive zeros of a nontrivial solution* $u(x)$ *of* (2.1) *and if*

(i) $$0 < P_1(x) \leq p_1(x)$$

(ii) $$P_o + \frac{q_o' - Q_o'}{2} + \frac{Q_o^2}{4P_1} \leq p_o$$

for $x \in [\alpha, \beta]$, *then every solution* $v(x)$ *of* (2.2) *has a zero in* $[\alpha, \beta]$.

Other special choices of $f(x)$ yield further variations along these lines.

3. An Oscillation Criterion

Equations of the form (2.1) can also be transformed into Riccati equations by means of the transformation $h(x) = -p_1u'/u$. Differentiating and making use of (2.1) one gets

$$h' = -p_o - q_oh + p_1h^2. \tag{2.4}$$

In order to generalize the techniques which are used in the case $q_o(x) \equiv 0$ we define

$$H(x) = h(x) - \frac{q_o(x)}{2p_1(x)}$$

which satisfies

$$H' = -p_o - \frac{q_o^2}{4p_1} - \left(\frac{q_o}{2p_1}\right)' + p_1H^2. \tag{2.5}$$

Since the singularities of $H(x)$ and $h(x)$ can be made to correspond, the argument in the proof of Theorem 1.3 can be applied to (2.5) to yield the following generalization of the Leighton oscillation criterion.

<u>2.3 Theorem</u>. <u>If</u> $\int_{\alpha}^{\infty} \frac{1}{p_1(x)}\, dx = \infty$ <u>and</u>

$$\lim_{c \to \infty} \left\{ \frac{q_o(c)}{2p_1(c)} + \int_{\alpha}^{c} \left[p_o(x) + q_o^2(x)/4p_1(x) \right] dx \right\} = \infty,$$

<u>then</u> (2.1) <u>is</u> <u>oscillatory</u> <u>at</u> ∞.

Other results for (1.1) based on Riccati equation arguments allow analogous generalizations to (2.1).

4. First Order Systems

Another generalization of the theory of Chapter 1 to nonselfadjoint problems is the consideration of first order systems of the form

$$\text{(2.6)} \qquad \begin{aligned} u' &= a(x)u + b(x)w \\ w' &= c(x)u + d(x)w \end{aligned}$$

$$\text{(2.7)} \qquad \begin{aligned} v' &= A(x)v + B(x)z \\ z' &= C(x)v + D(x)z \end{aligned}$$

where one need only require that the coefficients be continuous (or integrable) and in certain instances that $b(x) \geq 0$, $B(x) \geq 0$ on $\mathcal{J}$. In case $a(x) \equiv 0$, $b(x) = \frac{1}{p_1(x)}$, $c(x) = p_o(x)$ and $d(x) = q_o(x)/p_1(x)$, the system (2.6) reduces to (2.1); (2.7) allows an analogous reduction to (2.2). Thus it is of interest to outline an extension of the selfadjoint theory to (2.6) and (2.7).

Kamke [1], [2] has generalized the Prüfer transformation to this more general setting and used it to obtain comparison theorems for (2.6) and (2.7). Writing

$$u(x) = r(x) \sin \theta(x) \quad ; \quad w(x) = r(x) \cos \theta(x)$$

and substituting into (2.6) yields $\theta' = f(x,\theta)$ where

$$f(x,\theta) = b(x)\cos^2\theta + (a(x)-d(x))\sin\theta\cos\theta - c(x)\sin^2\theta.$$

Writing

$$v(x) = R(x)\sin\varphi(x) \quad ; \quad z(x) = R(x)\cos\varphi(x)$$

and substituting into (2.7) yields $\varphi' = F(x,\varphi)$ where

$$F(x,\varphi) = B(x)\cos^2\varphi + (A(x)-D(x))\sin\varphi\cos\varphi - C(x)\sin^2\varphi.$$

If $B(x) \geq b(x)$, $c(x) \geq C(x)$ and

$$4(B(x)-b(x))(c(x)-C(x)) \geq [(A(x)-D(x))-(a(x)-d(x))]^2$$

for $x \in [\alpha,\beta]$, then the quadratic form

$$(B(x)-b(x))\xi^2 + [(A(x)-D(x))-(a(x)-d(x))]\xi\eta + (c(x)-C(x))\eta^2$$

is nonnegative definite on $[\alpha,\beta]$ and $F(x,\theta) \geq f(x,\theta)$ for all $x \in [\alpha,\beta]$ and all θ. By the theory of differential inequalities, if $\theta' = f(x,\theta)$, $\varphi' = F(x,\varphi)$, and $\varphi(\alpha) \geq \theta(\alpha)$, then $\varphi(x) \geq \theta(x)$ for $\alpha \leq x \leq \beta$. These observations lead to the following comparison theorem.

2.4 Theorem. *Suppose* $u(x)$, $w(x)$ *and* $v(x)$, $z(x)$ *are nontrivial solutions of* (2.6) *and* (2.7) *respectively. If* α, β *are successive zeros of* $u(x)$ *and*

(i) $B(x) \geq b(x) > 0$

(ii) $c(x) \geq C(x)$

(iii) $(B(x)-b(x))(c(x)-C(x)) \geq \frac{1}{4}(A(x)+d(x)-a(x)-D(x))^2$

for $x \in [\alpha,\beta]$, *then* $v(x)$ *has a zero in* $[\alpha,\beta]$.

5. A Picone Identity for First Order Systems

The other techniques considered in Chapter 1 also generalize to this setting. If $u(x)$, $w(x)$ and $v(x)$, $z(x)$ are solutions of (2.6) and (2.7), respectively, then the analogue of the Picone identity is

$$\frac{d}{dx}\left[\frac{u}{v}(vw-uz)\right] = (c-C-g_1-g_2)u^2$$

(2.8) $$+ (b - b^2/B)w^2 + \left[(a+d)-(A+D)\frac{b}{B}\right]uw + g_1u^2$$

$$+ By^2 + (2a-A+D)uy + g_2u^2$$

where $y = \left(\frac{b}{B}w - \frac{zu}{v}\right)$. Choosing $g_1(x)$ and $g_2(x)$ so as to "complete the square" in the quadratic forms in u, w and u, y, one proceeds as in the proof of Theorems 1.2 and 2.1. The modifications to the proofs of these theorems are also valid for (2.8).

Finally we note that the system (2.6) also admits a transformation leading to a Riccati equation. Defining $h(x) = -w(x)/u(x)$ and making use of (2.6) yields

$$h'(x) = -c + (d-a)h + bh^2.$$

Defining $H(x) = h + \frac{d-a}{2b}$ one gets

$$H'(x) = -c - \frac{(d-a)^2}{4b^2} - \left(\frac{d-a}{2b}\right)' + bH^2$$

which corresponds to (1.6) in the selfadjoint case.

CHAPTER 3

PARTIAL DIFFERENTIAL EQUATIONS

1. A Comparison Theorem for Elliptic Equations

As suggested by the physical interpretations of Chapter 1, the Sturm-Picone theorem and much of the related theory should allow generalization to certain partial differential equations. The generalization to elliptic equations turns out to be the most satisfactory and complete, and we begin by generalizing (1.1) and (1.2) to the elliptic equations

$$\ell u \equiv - \sum_{i,j=1}^{n} \frac{\partial}{\partial x_j}\left(p_{ij}(x)\frac{\partial u}{\partial x_i}\right) + p_o(x)u = 0 \tag{3.1}$$

$$Lv \equiv - \sum_{i,j=1}^{n} \frac{\partial}{\partial x_j}\left(P_{ij}(x)\frac{\partial v}{\partial x_i}\right) + P_o(x)v = 0. \tag{3.2}$$

Here $x = (x_1,\cdots,x_n)$ represents a point in E^n and the P_{ij} and p_{ij} are assumed real valued and of class C^1 in an appropriate domain D while $p_o(x)$ and $P_o(x)$ are continuous. The ellipticity of (3.1) and (3.2) require that $p_{ij} = p_{ji}$, $P_{ij} = P_{ji}$, $\Sigma p_{ij}\xi_i\xi_j > 0$ and $\Sigma P_{ij}\xi_i\xi_j > 0$ for all real n-tuples $\underline{\xi} = (\xi_1,\cdots,\xi_n)$, $\underline{\xi} \neq \underline{0}$, and for all $x \in D$. Much of the theory remains valid for equations of elliptic-parabolic type where the quadratic forms generated by the p_{ij} and P_{ij} need only be positive semidefinite.

Numerous proofs and generalizations of the Sturm-Picone theorem to (3.1) and (3.2) have been presented, with most authors overlooking the original generalization due to M. Picone [2]. We shall later generalize Picone's technique to non-selfadjoint equations and present instead a proof due to the author [1] which is consistent with the physical motivation presented in Chapter 1. The notation $\frac{\partial u}{\partial \nu}$ is used to denote the transverse derivative $\frac{\partial u}{\partial \nu} \equiv \Sigma a_{ij}\frac{\partial u}{\partial x_i}\frac{\partial \nu}{\partial x_j}$ where $\frac{\partial \nu}{\partial x_j}$ is the cosine of the angle between x_j and the exterior normal.

<u>3.1 Theorem</u>. <u>Suppose</u> u(x) <u>and</u> v(x) <u>are nontrivial solutions of</u> (3.1) <u>and</u> (3.2), <u>respectively, in a bounded regular domain</u> D <u>and that</u>

(i) $0 < \Sigma P_{ij}(x)\xi_i\xi_j \leq \Sigma p_{ij}(x)\xi_i\xi_j$ <u>for all n-tuples</u> ξ <u>and all</u> $x \in D$,

(ii) $P_o(x) \leq p_o(x)$ <u>for all</u> $x \in D$.

<u>If</u> $\frac{\partial u}{\partial \nu} + \sigma(x)u = 0$ <u>and</u> $\frac{\partial v}{\partial \nu} + \tau(x)v = 0$ <u>on</u> ∂D <u>with</u> $-\infty < \tau(x) \leq \sigma(x) \leq +\infty$, <u>then either</u> v(x) <u>has a zero in the interior of</u> D <u>or else</u> u <u>is a constant multiple of</u> v.

<u>Proof</u>. Let $B_1 = \{x \in \partial D | \tau(x) < \infty\}$ and $B_2 = \{x \in \partial D | \sigma(x) < \infty\}$, so that $B_1 \supset B_2$. Without loss of generality we may assume $\iint_D u^2\, dx = 1$. If $v(x) \neq 0$ in D, then v(x) is the first eigenfunction of $Lv = \Lambda v$, $\frac{\partial v}{\partial \nu} + \tau v = 0$, corresponding to the eigenvalue $\Lambda_1 = 0$. Therefore by the variational characterization of Λ_1 we have for the class of "admissible" normalized $\varphi(x)$

$$0 = \inf \iint_D \left[\Sigma P_{ij} \frac{\partial \varphi}{\partial x_i} \frac{\partial \varphi}{\partial x_j} + P_o \varphi^2\right] dx + \int_{B_1} \tau\varphi^2\, d\bar{x}$$

$$\leq \iint_D \left[\Sigma P_{ij} \frac{\partial u}{\partial x_i} \frac{\partial u}{\partial x_j} + P_o u^2\right] dx + \int_{B_1} \tau u^2\, d\bar{x}$$

$$\leq \iint_D \left[\Sigma p_{ij} \frac{\partial u}{\partial x_i} \frac{\partial u}{\partial x_j} + p_o u^2\right] dx + \int_{B_2} \sigma u^2\, d\bar{x}.$$

However by (3.1) and Green's theorem the last term is zero and we have equality throughout. Therefore u(x) is an extremal function in the variational characterization of Λ_1 and therefore a constant multiple of v(x).

If $s(x) \equiv +\infty$ on ∂D, then D is a nodal domain for u(x) and we have the following.

<u>3.2 Corollary</u>. <u>If</u> D <u>is a nodal domain for a nontrivial solution</u> u(x) <u>of</u> (3.1)

and if (i) and (ii) of Theorem 3.1 are satisfied in D, then every solution $v(x)$ of (3.2) has a zero in D or else is a constant multiple of u.

2. A Picone Identity for Elliptic Equations

A more general result (Kreith [2], Dunninger [1]) can be obtained by means of a generalized Picone identity

$$\sum_j \frac{\partial}{\partial x_j}\left[\frac{u}{v}\left(v\sum_i p_{ij}\frac{\partial u}{\partial x_i} - u\sum_i P_{ij}\frac{\partial v}{\partial x_i}\right)\right] =$$

$$u\sum_{i,j}\frac{\partial}{\partial x_j}\left(p_{ij}\frac{\partial u}{\partial x_i}\right) - \frac{u^2}{v}\sum_{i,j}\frac{\partial}{\partial x_j}\left(P_{ij}\frac{\partial v}{\partial x_i}\right) \tag{3.3}$$

$$+ \sum_{i,j}(p_{ij}-P_{ij})\frac{\partial u}{\partial x_i}\frac{\partial u}{\partial x_j} + \sum_{i,j}P_{ij}\left(\frac{\partial u}{\partial x_i} - \frac{u}{v}\frac{\partial v}{\partial x_i}\right)\left(\frac{\partial u}{\partial x_j} - \frac{u}{v}\frac{\partial v}{\partial x_j}\right)$$

valid whenever u and v are sufficiently regular and $v(x) \neq 0$. In addition, for any differentiable vector valued function $\underline{f}(x) = (f_1(x),\cdots,f_n(x))$ we have

$$\nabla\cdot(u^2\underline{f}) = 2u\nabla u\cdot\underline{f} + u^2\nabla\cdot\underline{f}. \tag{3.4}$$

These identities can be used to establish a class of Sturmian comparison theorems for

$$-\sum_{i,j=1}^{n}\frac{\partial}{\partial x_j}\left(p_{ij}(x)\frac{\partial u}{\partial x_i}\right) + \sum_{i=1}^{n} q_i(x)\frac{\partial u}{\partial x_i} + p_o(x)u = 0, \tag{3.5}$$

$$-\sum_{i,j=1}^{n}\frac{\partial}{\partial x_j}\left(P_{ij}(x)\frac{\partial v}{\partial x_i}\right) + \sum_{i=1}^{n} Q_i(x)\frac{\partial u}{\partial x_i} + P_o(x)v = 0. \tag{3.6}$$

In order to formulate our results we shall require nonnegative functions $g_1(x)$ and $g_2(x)$ which assure that the quadratic forms

$$Q_1(\xi,\underline{f}) \equiv \sum_{i,j=1}^{n}(p_{ij}-P_{ij})\xi_i\xi_j + \xi_{n+1}\sum_{i=1}^{n}(q_i-Q_i+f_i)\xi_i + g_1\xi_{n+1}^2$$

and

$$Q_2(\xi) \equiv \sum_{i,j=1}^{n} P_{ij}\xi_i\xi_j + \xi_{n+1}\sum_{i=1}^{n} Q_i\xi_i + g_2\xi_{n+1}^2$$

are nonnegative definite.

3.3 Theorem. *If D is a nodal domain for a nontrivial solution* $u(x)$ *of* (3.5) *and there exists a differentiable vector function* $\underline{f}(x) = (f_1(x),\cdots,f_n(x))$ *such that*

$$\iint_D \left[\sum_{i,j}(p_{ij}-P_{ij})\frac{\partial u}{\partial x_i}\frac{\partial u}{\partial x_j} + u\sum_{i=1}^{n}\frac{\partial u}{\partial x_i}(q_i-Q_i+2f_i) + (p_o-P_o-g_2+\nabla\cdot\underline{f})u^2\right]dx > 0,$$

then every solution $v(x)$ *of* (3.6) *has a zero in* $\overline{D}$.

Proof. If $v(x) \neq 0$ in $\overline{D}$ then (3.3) is valid in D. Adding (3.3) and (3.4), making use of (3.5) and (3.6), and integrating over D yields

$$0 = \iint_D \left[\sum_{i,j}(p_{ij}-P_{ij})\frac{\partial u}{\partial x_i}\frac{\partial u}{\partial x_j} + u\sum_i\frac{\partial u}{\partial x_i}(q_i-Q_i+2f_i) + (p_o-P_o-g_2+\nabla\cdot\underline{f})u^2\right]dx$$

$$+ \iint_D \left[\sum_{i,j}P_{ij}\left(\frac{\partial u}{\partial x_i}-\frac{u}{v}\frac{\partial v}{\partial x_i}\right)\left(\frac{\partial u}{\partial x_j}-\frac{u}{v}\frac{\partial v}{\partial x_j}\right) + u\sum_i Q_i\left(\frac{\partial u}{\partial x_i}-\frac{u}{v}\frac{\partial v}{\partial x_i}\right) + g_2u^2\right]dx.$$

Since the second integral above is nonnegative by the definition of $g_2(x)$, the hypotheses assure the desired contradiction.

The choice $2f_i(x) = q_i(x) - Q_i(x)$ for $i=1,\cdots,n$ now yields the following.

3.4 Corollary. *If D is a nodal domain for a nontrivial solution* $u(x)$ *of* (3.5) *and*

$$\iint_D \left[\sum_{i,j}(p_{ij}-P_{ij})\frac{\partial u}{\partial x_i}\frac{\partial u}{\partial x_j} + \left(p_o-P_o-g_2+\frac{1}{2}\sum_{i=1}^{n}\frac{\partial}{\partial x_i}(q_i-Q_i)\right)u^2\right]dx > 0$$

then every solution $v(x)$ *of* (3.6) *has a zero in* $\overline{D}$.

The criterion of Corollary 3.4 was first established by Swanson [3].

The choice $f_i(x) \equiv 0$ for $i=1,\cdots,n$ yields the following.

3.5 Corollary. *If D is a nodal domain for a nontrivial solution u(x) of (3.5) and*

$$\iint_D (p_o-P_o-g_1-g_2)u^2\, dx > 0,$$

then every solution v(x) of (3.6) has a zero in $\overline{D}$.

These comparison theorems for nonselfadjoint equations are "weak" in the sense that they only establish the existence of a zero of v(x) in the closure of D and do not yield the stronger results of Theorem 3.1 and its Corollary. Comparison theorems of this stronger variety have recently been obtained by Allegretto [1] who studies generalized solutions of elliptic equations. With these techniques he is able to replace the strict inequality in the hypotheses of Theorem 3.3 and its corollaries by inequality and establish the stronger result that v changes sign in D or else is of the form $e^w u$ for an appropriate function w(x). This generalizes an analogous result for ordinary differential equations due to Swanson [1].

3. Oscillation Theory for Elliptic Equations

Given the above Sturmian comparison theorem for elliptic equations one can establish oscillation criteria in a direct way. Writing $|x| = \sqrt{x_1^2+\ldots+x_n^2}$ and defining $E_R = \{x \in E^n \,\big|\, |x| > R\}$, the equation (3.1) is said to be *nodally oscillatory* at $|x| = \infty$ if for every $R > 0$ (3.1) has a nodal domain in E_R. If (3.1) allows separation of variables in some appropriate coordinate system, one can use oscillation criteria for ordinary differential equations to establish that (3.1) is nodally oscillatory at ∞. Applying Theorem 3.1, one concludes that for a class of more general equations of the form (3.2) every global solution has arbitrarily large zeros. This technique has been used by the author [3], Headley and Swanson [1], and elsewhere.

The question naturally arises whether there are more sensitive oscillation criteria for (3.2) which are analogous to those of Theorem 1.3 insofar as they involve improper integrals of the coefficients and do not require comparison with a simpler oscillatory equation. A number of techniques for arriving at such integral criteria have recently been developed by Kreith and Travis [1], Noussair [1], and Swanson [2]. The following two theorems are based on Swanson's technique. Letting (r,θ) denote hyperspherical coordinates in E^n, we define $N(r,\theta)$ to the largest eigenvalue of the symmetric matrix $P_{ij}(x)$ and define

$$\pi(r) = \int_{\Omega} N(r,\theta)\, d\theta$$

$$\gamma(r) = \int_{\Omega} P_o(r,\theta)\, d\theta$$

where Ω denotes the (n-1)-sphere in E^n.

3.6 Theorem. *If the ordinary differential equation*

$$-\frac{d}{dr}\left(r^{n-1}\pi(r)\frac{dw}{dr}\right) + r^{n-1}\gamma(r)w = 0 \tag{3.7}$$

is oscillatory at $r = \infty$, *then* (3.2) *is nodally oscillatory at* $|x| = \infty$.

Proof. Let $w(r)$ be a nontrivial solution of (3.7) with zeros at $r_1 < r_2 < \cdots < r_k < \cdots$, where $r_k \uparrow \infty$. Given $R > 0$ there exists k such that $D_k = \{x | r_k < |x| < r_{k+1}\} \subset E_R$. Defining $v(x) = w(|x|)$ we have $v(x) = 0$ on ∂D_k. Consider now the eigenvalue problem

$$Lv \equiv -\sum_{i,j} \frac{\partial}{\partial x_j}\left(P_{ij}\frac{\partial v}{\partial x_i}\right) + P_o v = \Lambda v \qquad \text{in } D_k$$

$$v = 0 \qquad \text{on } \partial D_k.$$

If Φ denotes the class of "admissible" functions for this eigenvalue problem, the classical variational characterization yields

$$\Lambda_1 = \min_{\varphi\in\Phi} \iint_{D_k} \left[\Sigma P_{ij} \frac{\partial\varphi}{\partial x_i}\frac{\partial\varphi}{\partial x_j} + P_o\varphi^2\right] dx$$

$$\leq \iint_{D_k} \left[\Sigma P_{ij} \frac{\partial v}{\partial x_i}\frac{\partial v}{\partial x_j} + P_o v^2\right] dx$$

$$\leq \int_{r_k}^{r_{k+1}} r^{n-1} \left\{\int_\Omega \left[N\Sigma \left(\frac{\partial v}{\partial x_i}\right)^2 + P_o v^2\right] d\theta\right\} dr$$

$$= \int_{r_k}^{r_{k+1}} r^{n-1} \left[\pi \left(\frac{dw}{dr}\right)^2 + \gamma w^2\right] dr = 0.$$

Thus $\Lambda_1 \leq 0$, and by classical variational principles there is a subdomain $D' \subset D_k$ in which the above eigenvalue problem has $\Lambda_1' = 0$. Clearly D' is the desired nodal domain.

As an example of how this theorem yields integral oscillation criteria we consider the equation $-\Delta v + P_o(x,y)v = 0$ in E^2.

3.7 Theorem. *If*

$$\iint_{E^2} P_o(x,y)\, dx\, dy = -\infty$$

then $-\Delta v + P_o v = 0$ *is nodally oscillatory at* $|x| = \infty$.

Proof. Defining $\gamma(r) = \int_0^{2\pi} P_o(r,\theta)\, d\theta$, it is sufficient by Theorem 3.6 to show that $-\frac{d}{dr}\left(r\frac{dw}{dr}\right) + r\gamma(r)w = 0$ is oscillatory at $r = \infty$. By Theorem 1.3 this is the case if $\int_0^\infty r\gamma(r)\, dr = -\infty$ -- i.e. if $\int_0^\infty r\left[\int_0^{2\pi} P_o(r,\theta)\, d\theta\right] dr = -\infty$.

As an alternate approach to results such as the above, one might seek to generalize the Riccati equation methods used in the proof of Theorem 1.3. Given a solution $u(x)$ of (3.1), one can indeed use the substitution

$$h_j(x) = -\frac{1}{u}\sum_i P_{ij}\frac{\partial u}{\partial x_i} \quad ; \quad j=1,\cdots,n$$

to obtain an equation of Riccati type

$$\nabla\cdot\underline{h} = -p_o + \underline{h}(p_{ij})^{-1}\underline{h}^T$$

where $\underline{h} = (h_1,\cdots,h_n)$ and T denotes transpose. While this transformation has been used successfully by Wong [1] to establish Wirtinger type inequalities, it does not allow for a full generalization due to the complexity of the vector formula for $\nabla(\underline{h}\cdot\underline{h}^T)$ or more generally for $\nabla(\underline{h}(p_{ij})^{-1}\underline{h}^T)$.

4. Hyperbolic Initial Boundary Value Problems

Another possible generalization (Kreith [4]) of the results of Chapter 1 is to equations of hyperbolic type. Motivated by the simple harmonic motion interpretation of the Sturm comparison theorem, we consider a pair of hyperbolic equations of the form

$$u_{tt} - u_{xx} + f(x,t) = 0 \tag{3.8}$$

$$v_{tt} - v_{xx} + F(x,t) = 0 \tag{3.9}$$

representing the motion of two vibrating strings of unit density and elastic constant which are oscillating about the equilibrium lines $u = 0$ and $v = 0$, subject to continuous restoring forces $f(x,t)$ and $F(x,t)$, respectively. It is reasonable to expect that if $F(x,t) \geq f(x,t)$, then (3.9) should in some sense oscillate more rapidly than (3.8). However experimentation with simple examples allowing separation of variables shows this is not the case without some auxiliary condition. In physical terms it is necessary to consider finite strings which are elastically bound at the ends, with the string which is to oscillate faster being more tightly bound. In mathematical terms it is possible to establish an analogue of the Sturm comparison theorem for hyperbolic initial boundary value problems.

3.8 Theorem. *Let $u(x,t)$ be a solution of (3.8) which is positive on* $R = \{(x,t)|\alpha < x < \beta;\ t_o < t < T\}$ *which satisfies*

(i) $\qquad u(x,t_o) = u(x,T) = 0 \quad ; \quad \alpha \leq x \leq \beta$

(ii) $\qquad u_x(\alpha,t) - \sigma_1(t)u(\alpha,t) = 0 \quad ; \quad t_o \leq t \leq T$

(iii) $\qquad u_x(\beta,t) + \sigma_2(t)u(\beta,t) = 0 \quad ; \quad t_o \leq t \leq T.$

If $F(x,t) \geq f(x,t)$ *on* R *and* $-\infty < \sigma_i(t) \leq \tau_i(t) \leq +\infty$ *for* $i=1,2$ *and* $t_o \leq t \leq T$, *then every solution of* (3.9) *satisfying*

(iv) $\qquad v_x(\alpha,t) - \tau_1(t)v(\alpha,t) = 0$

(v) $\qquad v_x(\beta,t) + \tau_2(t)v(\beta,t) = 0$

has a zero in $\overline{R}$.

Proof. Suppose to the contrary that $v(x) > 0$ in $\overline{R}$. Multiplying through (3.8) and (3.9) by v and u, respectively, and subtracting we get

$$(vu_t - uv_t)_t - (vu_x - uv_x)_x = (F-f)uv.$$

Integrating over R and applying Green's theorem yields

$$\int_{\partial D} (vu_t - uv_t)\, dx + (vu_x - uv_x)\, dt = -\iint_D (F-f)uv\, dx\, dt. \tag{3.10}$$

Using the boundary conditions on u and v the above boundary integral becomes

$$\int_{(\alpha,t_o)}^{(\beta,t_o)} vu_t\, dx + \int_{(\beta,t_o)}^{(\beta,T)} (\tau_2 - \sigma_2)uv\, dt - \int_{(\alpha,T)}^{(\beta,T)} vu_t\, dx - \int_{(\alpha,t_o)}^{(\alpha,T)} (\sigma_1 - \tau_1)uv\, dt.$$

Our hypotheses assume that the above expression is positive, contradicting the non-positivity of the right side of (3.10).

Remarks.

1. Theorem 3.8 is an analogue of the Sturm comparison theorem in that the method of proof requires (3.8) and (3.9) to be selfadjoint and to have the same principal part. It would be of considerable interest to find an analogue of the Sturm-Picone comparison theorem for hyperbolic equations, thereby allowing comparison of equations with different principal parts and perhaps nonselfadjoint equations.

2. The hypotheses of Theorem 3.8 can clearly be strengthened to assure that $v(x)$ have a zero in R rather than $\bar{R}$. The technique can also be applied to differential inequalities.

5. Hyperbolic Characteristic Initial Value Problems

Theorem 3.8 has a direct analogue for a class of characteristic initial value problems. Consider the equations

$$u_{xy} + f(x,y)u = 0 \tag{3.11}$$

$$v_{xy} + F(x,y)v = 0 \tag{3.12}$$

in a domain D enclosed by two nonintersecting noncharacteristic curves C_{r_1} and C_{r_2} whose graphs are given in the first quadrant by

$$y = g_{r_i}(x) \quad , \quad 0 < x < \infty \quad ; \quad i=1,2$$

where each $g_{r_i}(x)$ is continuously differentiable, strictly decreasing, and satisfies

$$\lim_{x \to 0} g_{r_i}(x) = \infty \quad ; \quad \lim_{x \to \infty} g_{r_i}(x) = 0 \quad ; \quad i=1,2.$$

We say that $v(x,y)$ is u-regular with respect to D if

$$\lim_{x \to \infty} \int_{g_{r_1}(x)}^{g_{r_2}(x)} |u(x,y)v_y(x,y)| \, dy = 0$$

$$\lim_{y \to \infty} \int_{g_{r_1}^{-1}(y)}^{g_{r_2}^{-1}(y)} |v(x,y)u_x(x,y)| \, dx = 0.$$

3.9 Theorem. *Let $u(x,y)$ be a solution of (3) which vanishes on C_{r_1} and C_{r_2} and is positive in D. If $F(x,y) \geq f(x,y)$, $F(x,y) \not\equiv f(x,y)$ in D, then every u-regular solution v of (3.12) has a zero in D.*

<u>Proof</u>. Suppose v is positive in D. Multiplying (3.11) and (3.12) by v and u, respectively, and subtracting we get

$$(vu_x)_y - (uv_y)_x = (F-f)uv.$$

Let $D_M = \{(x,y) \in D \mid 0 < x < M;\ 0 < y < M\}$. Integrating the above over D_M and applying Green's theorem yields

$$\int_{\partial D_M} vu_x\, dx + uv_y\, dy = -\iint_{D_M} (F-f)uv\, dx\, dy.$$

Defining

$$x_i = g_{r_i}(M) \quad ; \quad y_i = g_{r_i}^{-1}(M) \quad , \quad i=1,2$$

the above boundary integral can be written

$$\int_{x_1}^{M} v(x,g_{r_1}(x))u_x(x,g_{r_1}(x))\, dx + \int_{y_1}^{y_2} u(M,y)v_y(M,y)\, dy$$

$$- \int_{x_2}^{M} v(x,g_{r_2}(x))\, dx - \int_{x_1}^{x_2} v(x,M)u_x(x,M)\, dx.$$

Since v is assumed u-regular in D, the second and fourth integrals tend to zero as $M \to \infty$. Since $u_x \geq 0$ on C_{r_1} and $u_x \leq 0$ on C_{r_2}, the first and third terms make nonnegative contributions for all M, and this fact contradicts the fact that

$$-\iint_{D_M} (F-f)uv\, dx\, dy < 0.$$

The question remains whether there is a reasonable class of u-regular solutions of (3.12). It can be shown (Kreith [5; Theorem 4] that for the important case $f(x,y)$ is of the form $\lambda(xy)^{\alpha}$ with $\lambda > 0$ and $\alpha > -1$ and $F(x,y)$ bounded, the following condition assures u-regularity: there exists $\delta > 0$ such that

$$\lim_{x \to \infty} v(x,y) = 0 \quad \text{uniformly for } 0 \leq y \leq \delta,$$

$$\lim_{x \to \infty} v(x,y) = 0 \quad \text{uniformly for } 0 \leq x \leq \delta.$$

Since we are free to assign the values of $v(x,y)$ along the characteristics $x = 0$ and $y = 0$, $v(x,y)$ will be u-regular if $\lim_{x \to \infty} v(x,0) = \lim_{y \to \infty} v(0,y) = 0$ and $v(x,y)$ is sufficiently well behaved in a strip near these characteristics.

6. Oscillation Theory for Hyperbolic Equations

The comparison theorems of the preceding section can be used to establish oscillation criteria for (3.9) and (3.12), respectively, by considering special solutions of (3.8) and (3.11). Rather than stating the oscillation criteria in detail, we shall simply exhibit some appropriate solutions of the comparison equations (3.8) and (3.11).

For initial boundary value problems we consider

$$\begin{aligned} u_{tt} - u_{xx} + pu &= 0 \\ u_x(\alpha,t) - \sigma_1 u(\alpha,t) &= 0 \\ u_x(\beta,t) + \sigma_2 u(\beta,t) &= 0 \end{aligned} \tag{3.13}$$

where p is a function of t only and σ_1, σ_2 are constants. If λ_1 is the first eigenvalue of

$$-\frac{d^2y}{dx^2} = \lambda y$$

$$y'(\alpha) - \sigma_1 y(\alpha) = 0$$

$$y'(B) + \sigma_2 y(\beta) = 0$$

and if $\frac{d^2T}{dt^2} + (p+\lambda_1)T = 0$ is oscillatory at $t = \infty$, then separation of variables shows that (3.13) possesses a sequence of nodal domains of the form

$D_k = \{(x,t) | \alpha < x < \beta,\ t_k < t < t_{k+1}\}$ where $t_k \uparrow \infty$. These can be used to establish oscillation criteria for a class of equations of the form (3.9).

For characteristic initial value problems we consider

$$u_{xy} + f(x,y)u = 0 \tag{3.14}$$
$$u(x,0) = u(0,y) = 1$$

where $f(x,y)$ has the special form $k(xy)^{\gamma}$ with $k > 0$ and $\gamma > -1$. The problem (3.14) has a solution of the form $u(x,y) = u(z)$, where $z = xy$ and

$$z\frac{d^2u}{dz^2} + \frac{du}{dz} + kz^{\gamma}u = 0 \tag{3.15}$$
$$u(0) = 1.$$

The initial value problem (3.15) has the solution $u(z) = J_o\left(\frac{2\sqrt{k}}{\gamma+1} z^{\frac{\gamma+1}{2}}\right)$. Thus (3.14) has a sequence of nodal domains enclosed by the noncharacteristic curves C_{r_n} determined by

$$xy = \left(\frac{\gamma+1}{2\sqrt{k}} j_n\right)^{\frac{2}{\gamma+1}}$$

where j_n is the n-th zero of the Bessel function $J_o(z)$. Using these nodal domains one can establish oscillation criteria for u-regular solutions of (3.12).

CHAPTER 4

RELATED TOPICS IN ANALYSIS

1. Calculus of Variations

Oscillation theory is intimately connected with the calculus of variations. Specifically, a special case of the Picone identity (1.4) yields the extended Legendre condition, and a variational principle closely related to the Legendre condition can be used to establish comparison theorems of a somewhat more general nature than the Sturm-Picone theorem.

Given a sufficiently regular functional of the form

$$F[y] = \int_{\alpha}^{\beta} f(x,y,y')\, dx$$

we consider the classical problem of finding sufficient conditions for such a functional to have a minimum in the class of continuously differentiable functions satisfying $y(\alpha) = y(\beta) = 0$. A necessary condition for $F[y_o]$ to have a minimum is that the first variation $\delta F[y]$ be zero at y_o; a sufficient condition is that in addition the second variation $\delta^2 F[y]$ be positive at y_o (see Gelfand and Fomin [1]). The implications of these requirements follow from using Taylor's theorem to write

$$\begin{aligned}\Delta F[y] &= F[y+\eta] - F[y] \\ &= \int_{\alpha}^{\beta} [f_y(x,y,y')\eta + f_{y'}(x,y,y')\eta']\, dx \\ &+ \frac{1}{2}\int_{\alpha}^{\beta} [f_{yy}\eta^2 + 2f_{yy'}\eta\eta' + f_{y'y'}\eta'^2]\, dx \\ &+ \text{terms of higher order,}\end{aligned}$$

where η is of class C^1 on $[\alpha,\beta]$ and satisfies $\eta(\alpha) = \eta(\beta) = 0$. The first integral above is $\delta F[y]$, and its vanishing for all admissible η leads to the Euler equation

$$f_y - \frac{d}{dx} f_{y'} = 0.$$

The second integral above is $\delta^2 F[y]$, and a necessary condition that $\delta^2 F[y_o] > 0$ for all admissible $\eta \not\equiv 0$ is that $f_{y'y'}(x,y_o(x),y_o'(x)) \geq 0$ for $\alpha \leq x \leq \beta$ (the so-called Legendre necessary condition). In order to obtain sufficient conditions we perform an integration by parts to write

$$\delta^2 F[y] = \int_\alpha^\beta \left[f_{y'y'} \eta'^2 + \left(f_{yy} - \frac{d}{dx} f_{yy'} \right) \eta^2 \right] dx$$

and set

$$P_1(x) = f_{y'y'}(x,y(x),y'(x))$$

$$P_o(x) = f_{yy}(x,y(x),y'(x)) - \frac{d}{dx} f_{yy'}(x,y(x),y'(x)).$$

The basic connection between oscillation theory and the calculus of variations is contained in the following.

4.1 Theorem. *Suppose* $P_1(x) > 0$ *for* $\alpha \leq x \leq \beta$. *If the differential equation*

$$-(P_1(x)v')' + P_o(x)v = 0 \tag{4.1}$$

is disconjugate on $[\alpha,\beta]$, *then*

$$\int_\alpha^\beta (P_1\eta'^2 + P_o\eta^2)\, dx > 0 \tag{4.2}$$

for all admissible $\eta(x) \not\equiv 0$.

Proof. If (4.1) is disconjugate on $[\alpha,\beta]$, then there exists a solution $v(x)$ of (4.1) which is different from zero on $[\alpha,\beta]$. The Picone identity (1.4) with $p_1(x) \equiv P_1(x)$ yields

$$\frac{d}{dx}\left[\eta P_1 \eta' - \eta^2 \frac{P_1 v'}{v} \right] = \eta(P_1\eta')' - \frac{\eta^2}{v}(P_1 v')' + P_1\left(\eta' - \frac{\eta}{v} v'\right)^2. \tag{4.3}$$

Making use of (4.1) and cancelling $\eta(P_1\eta')'$ yields

(4.4)
$$P_1\eta'^2 - \frac{d}{dx}\left[\eta^2 \frac{P_1 v'}{v}\right] = -P_o\eta^2 + P_1(\eta' - \frac{\eta}{v} v')^2,$$

so that

$$0 \leq \int_\alpha^\beta P_1(\eta' - \eta \frac{v'}{v})^2\, dx = \int_\alpha^\beta (P_1\eta'^2 + P_o\eta^2)\, dx$$

with equality if and only if $\eta' \equiv \eta \frac{v'}{v}$ -- i.e. if and only if η is a multiple of v. But since $\eta(\alpha) = \eta(\beta) = 0$ and $v(x)$ is assumed positive on $[\alpha,\beta]$ equality is precluded and (4.2) is established for all admissible $\eta(x) \not\equiv 0$.

4.2 Corollary. *If there exists an admissible* $\eta(x) \not\equiv 0$ *such that*

(4.5)
$$\int_\alpha^\beta (P_1\eta'^2 + P_o\eta^2)\, dx \leq 0,$$

then every solution of (4.1) *has a zero in* $[\alpha,\beta)$.

Leighton [2], [3] has noted that the above Corollary leads to an alternate proof and a generalization of the Sturm-Picone theorem. Rewriting (4.5) in the form

(4.6)
$$\int_\alpha^\beta [(p_1-P_1)\eta_1'^2 + (p_o-P_o)\eta^2]\, dx \geq \int_\alpha^\beta (p_1\eta'^2 + p_o\eta^2)\, dx$$

we obtain the following.

4.3 Corollary. *If there exists an admissible* η *such that* (4.6) *is satisfied, then every solution of* (4.1) *has a zero in* $[\alpha,\beta)$.

If $\eta(x)$ also satisfies the equation

(4.7)
$$-(p_1u')' + p_ou = 0,$$

then the right side of (4.6) is zero and the Corollary reduces essentially to the Sturm-Picone theorem.

The technique outlined above generalizes readily to any situation where a Picone-type inequality is known: *a special case of the Picone identity yields a variational principle analogous to that of Corollary 4.2, and this variational* principle can in turn be used to establish a Sturmian comparison theorem. In the case of selfadjoint elliptic equations this approach was used by Clark and Swanson [1]. Setting $p_{ij}(x) \equiv P_{ij}(x)$ in the Picone identity (3.3), making use of the operator L defined by

$$Lv \equiv -\Sigma \frac{\partial}{\partial x_j}\left(P_{ij}\frac{\partial v}{\partial x_i}\right) + P_o v,$$

and cancelling the terms $u\Sigma \frac{\partial}{\partial x_j}\left(P_{ij}\frac{\partial u}{\partial x_i}\right)$ yields

$$\begin{aligned}\sum_{i,j} P_{ij}\frac{\partial u}{\partial x_i}\frac{\partial u}{\partial x_j} - \sum_j \frac{\partial}{\partial x_j}\left(\frac{u^2}{v}\sum_i P_{ij}\frac{\partial v}{\partial x_i}\right) &= \frac{u^2}{v} Lv \\ &- P_o u^2 + \sum_{i,j} P_{ij}\left(\frac{\partial u}{\partial x_i} - \frac{u}{v}\frac{\partial v}{\partial x_i}\right)\left(\frac{\partial u}{\partial x_j} - \frac{u}{v}\frac{\partial v}{\partial x_j}\right)\end{aligned} \tag{4.8}$$

which is the generalization of (4.3) to elliptic equations. Swanson [3] has also used an identity of this form to establish comparison theorems for nonselfadjoint elliptic equations and to generalize these results to differential inequalities and to certain kinds of singular problems [4], [5].

2. Bounds for Eigenvalues

A connection between oscillation theory and eigenvalue problems was already exhibited in the proof of Theorem 3.1. Another example of this connection was given in Kreith [6] where the Picone identity (1.4) is used to establish upper and lower bounds for eigenvalues of elliptic operators of the form

$$\ell w \equiv -\sum_{i,j=1}^{n} \frac{\partial}{\partial x_j}\left(p_{ij}\frac{\partial w}{\partial x_i}\right) + p_o w. \tag{4.9}$$

4.4 Theorem. *Let $u(x)$ and $v(x)$ be positive and of class C^2 in a bounded domain D and satisfy the boundary conditions*

$$\text{(i)} \qquad \frac{\partial u}{\partial \nu} + \sigma u = 0 \qquad \underline{\text{on}}\ \partial D$$

$$\text{(ii)} \qquad \frac{\partial v}{\partial \nu} + \tau v = 0 \qquad \underline{\text{on}}\ \partial D,$$

respectively, with $-\infty < \tau(x) \leq \sigma(x) \leq \infty$. *Let* λ_1 *and* Λ_1 *be the first eigenvalues of* $\ell w = \lambda\rho(x)w$ *with the boundary conditions* (i) *and* (ii), *respectively. Then*

$$\Lambda_1 \leq \sup_{x \in D} \frac{\ell u}{\rho u}$$

and

$$\lambda_1 \geq \inf_{x \in D} \frac{\ell v}{\rho v}\ .$$

Proof. Since $v(x) \neq 0$, the Picone identity (3.3) is valid. Integrating (3.3) over D and applying Green's theorem, yields

$$\int_{\partial D} u^2 \left(\frac{1}{u}\frac{\partial u}{\partial \nu} - \frac{1}{v}\frac{\partial v}{\partial \nu}\right) d\bar{x} \geq -\iint_D \left[u\ \ell u - \frac{u^2}{v}\ \ell v\right] dx.$$

If $\ell v = \Lambda_1 \rho v$, then

$$0 \geq \int_{\partial D} u^2(\tau-\sigma)\ d\bar{x} \geq -\iint_D \rho u^2 \left(\frac{\ell u}{\rho u} - \Lambda_1\right) dx$$

so that $\Lambda_1 \leq \sup_{x \in D} \frac{\ell u}{\rho u}$. If $\ell u = \lambda_1 \rho u$, then

$$0 \geq \int_{\partial D} u^2(\tau-\sigma) \geq -\iint_D \rho u^2 \left(\lambda_1 - \frac{\ell v}{\rho v}\right) dx$$

so that $\lambda_1 \geq \inf_{x \in D} \frac{\ell v}{\rho v}$.

For the special case $\ell = -\Delta$, inequalities such as these were derived by Barta [1]. Swanson [4] has also used his identity (4.8) to obtain lower bounds for eigenvalues in a more general setting. Protter and Weinberger [1] have used maximum principles to obtain such estimates for nonselfadjoint equations.

3. Green's Functions

Let the elliptic operator ℓ be defined by

$$\ell u \equiv - \sum_{i,j=1}^{n} \frac{\partial}{\partial x_j}\left(p_{ij}(x) \frac{\partial u}{\partial x_i}\right) + p_o u$$

in a domain D. Under suitable assumptions regarding the regularity of the coefficients and the domain, it is possible to solve the boundary value problem

$$(4.10) \qquad \begin{aligned} \ell u &= f(x) && \text{in } D \\ \frac{\partial u}{\partial \nu} + \sigma u &= 0 && \text{on } \partial D \end{aligned}$$

by means of an integral operator

$$u(x) = \iint_D g(x,\xi) f(\xi)\, d\xi$$

where $g(x,\xi)$ is the Green's function associated with the operator ℓ and the boundary conditions of (4.10). As is well known, $g(x,\xi)$ is characterized by the fact that it is symmetric in x and ξ, satisfies $\ell g = 0$ and the boundary condition of (4.10) for $x \neq \xi$, and has an appropriate singularity at $x = \xi$ which implies that

$$\lim_{x \to \xi_o} g(x,\xi_o) = \lim_{\xi \to x_o} g(x_o,\xi) = +\infty$$

for all x_o and ξ_o in D.

It is frequently of interest to know whether $g(x,\xi)$ is nonnegative in $D \times D$, and comparison theorems can be used to establish criteria for the nonnegativity of the Green's function.

__4.5 Theorem__. __If the homogeneous problem__

$$(4.11) \qquad \begin{aligned} \ell v &= 0 \\ \frac{\partial v}{\partial \nu} + \sigma v &= 0 \end{aligned}$$

__has a solution__ $v(x)$ __which is positive in__ D, __then the Green's function associated with__ (4.10) __is nonnegative in__ $D \times D$.

Proof. Suppose to the contrary that $g(x_o,\xi_o) < 0$. Since $\lim_{x \to \xi_o} g(x,\xi_o) = +\infty$, there exists a proper subdomain $D_o \subset D$ such that $\xi_o \notin D_o$ and

$$g(x,\xi_o) < 0 \qquad \text{for } x \in D_o$$

$$g(x,\xi_o) = 0 \qquad \text{for } x \in \partial D_o \cap D$$

$$\frac{\partial g}{\partial \nu} + \sigma g = 0 \qquad \text{for } x \in \partial D_o \cap \partial D.$$

Applying Theorem 3.1 we conclude that any solution of (4.11) must have a zero in D_o, and this is the desired contradiction.

Two familiar criteria for nonnegative Green's functions follow readily from Theorem 4.5.

4.6 Corollary. *If the first eigenvalue of*

$$\ell u = \lambda u \qquad \text{in } D$$

$$\frac{\partial u}{\partial \nu} + \sigma u = 0 \qquad \text{on } \partial D$$

is positive, then the Green's function associated with (4.10) *is nonnegative in* $D \times D$.

4.7 Corollary. *If* $p_o(x) \geq 0$ *in* D *then the Green's function associated with* (4.10) *is nonnegative in* $D \times D$.

Using more general comparison theorems it is possible to establish similar criteria for the Green's function of nonselfadjoint problems to be nonnegative. An application of the Hopf maximum principle for elliptic equations can also be used to show that if $g(x_o,\xi_o) = 0$, then $g(x,\xi_o)$ changes sign at x_o; thus the above criteria actually establish the *positivity* of the Green's function in $D \times D$ (see Kreith [6], [7]).

4. An Ordering of Operators

The proof of Theorem 3.1 suggests that an ordering among differential operators is intimately connected with the oscillation properties under study. Specifically, it was the fact that "$L < \ell$" in the sense that the first eigenvalue of

$$Lv \equiv -\Sigma \frac{\partial}{\partial x_j}\left(P_{ij}\frac{\partial v}{\partial x_i}\right) + P_o v = \Lambda v \qquad \text{in } D$$

$$\frac{\partial v}{\partial \nu} + \tau v = 0 \qquad \text{on } \partial D$$

is smaller than the first eigenvalue of

$$\ell u \equiv -\Sigma \frac{\partial}{\partial x_j}\left(p_{ij}\frac{\partial u}{\partial x_i}\right) + p_o u = \lambda u \qquad \text{in } D$$

$$\frac{\partial u}{\partial \nu} + \sigma u = 0 \qquad \text{on } \partial D$$

which, together with certain properties of the eigenfunctions of ℓ and L, led to the conclusion that nontrivial solutions of $Lv = 0$ change sign in D.

The question arises whether any of these order considerations can be extended to nonselfadjoint equations, with affirmative answers having been given by Kreith [8], and Allegretto [1], [2]. Underlying such an operator ordering for nonselfadjoint operators of the form

$$\ell u \equiv -\Sigma \frac{\partial}{\partial x_j}\left(p_{ij}\frac{\partial u}{\partial x_i}\right) + \Sigma q_i \frac{\partial u}{\partial x_i} + p_o u \qquad \text{for } x \in D, \tag{4.12}$$

is that if the operator ℓ, the bounded domain D, and the appropriate boundary conditions are sufficiently regular for a Green's function to exist, then for a sufficiently large constant γ the Green's function $g_\gamma(x,\xi)$ associated with $\ell u + \gamma u$ is positive in $D \times D$. A generalization of the Perron-Frobenius theory of positive matrices due to Jentsch [1] and Rutman [1] (see also Krasnoselskii [1]) asserts that the integral operator

$$(\ell+\gamma I)^{-1}w \equiv \iint_D g(x,\xi)w(\xi)\, d\xi$$

has a positive simple eigenvalue μ, which is larger than the absolute value of any other eigenvalue and that the corresponding eigenfunction is positive in D

while all other eigenfunctions change sign in D. By the spectral mapping theorem it follows that the operator ℓ also has a distinguished real simple eigenvalue $\lambda_1 = \frac{1}{\mu_1} - \gamma$ which corresponds to a positive eigenfunction, while all other eigenvalues correspond to eigenfunctions which change sign. This fact motivates the abstract comparison theory to be presented in Chapter 5; it also leads to the operator ordering to be considered below.

Let λ_1 be the distinguished eigenvalue of $\ell u = \lambda u$ where ℓ is given by (4.12) in D and u = 0 on ∂D. We also consider the selfadjoint operator $L = \frac{1}{2}(\ell + \ell^*)$ which is defined by

$$\text{(4.13)} \qquad Lv \equiv -\Sigma \frac{\partial}{\partial x_j}\left(p_{ij} \frac{\partial v}{\partial x_i}\right) + \left(p_o - \frac{1}{2} \Sigma \frac{\partial q_i}{\partial x_i}\right)v \quad \text{for } x \in D,$$

and denote by Λ_1 the first eigenvalue of $Lv = \Lambda v$ with v = 0 on ∂D.

<u>4.8 Theorem.</u> <u>If</u> u(x) <u>is a solution of</u> $\ell u = 0$ <u>satisfying</u> $u \not\equiv 0$ <u>in</u> D <u>and</u> u = 0 <u>on</u> ∂D, <u>then every solution of</u> Lv = 0 <u>has a zero in</u> $\bar{D}$.

<u>Proof</u>. Since $\ell u = 0$ in D and u = 0 on ∂D, an application of Green's theorem yields

$$0 = \iint_D u\, \ell u\, dx = \iint_D \left[\Sigma p_{ij} \frac{\partial u}{\partial x_i} \frac{\partial u}{\partial x_j} + \left(p_o - \frac{1}{2} \Sigma \frac{\partial q_i}{\partial x_i}\right) u^2\right] dx.$$

From the elliptic analogue of Corollary 4.2 (see the remarks following Corollary 4.3) it follows that every solution of Lv = 0 has a zero in $\bar{D}$.

<u>4.9 Theorem.</u> <u>With</u> λ_1 <u>and</u> Λ_1 <u>defined as above</u>, $\Lambda_1 \leq \lambda_1$.

<u>Proof</u>. By classical variational theory,

$$\Lambda_1 = \inf_{\varphi \in \Phi} \iint_D \left[\Sigma p_{ij} \frac{\partial \varphi}{\partial x_i} \frac{\partial \varphi}{\partial x_j} + \left(p_o - \frac{1}{2} \Sigma \frac{\partial q_i}{\partial x_i}\right) \varphi^2\right] dx$$

where Φ consists of all normalized piecewise C^1 functions which vanish on ∂D. Since the normalized eigenfunction corresponding to λ_1 belongs to Φ we have by Green's theorem

$$\Lambda_1 \le \iint_D \left[\Sigma p_{ij} \frac{\partial u}{\partial x_i}\frac{\partial u}{\partial x_j} + \left(p_o - \frac{1}{2}\Sigma \frac{\partial q_i}{\partial x_i}\right)u^2\right] dx = \iint_D u\, \ell u = \lambda_1.$$

The preceding two theorems show that "$\frac{1}{2}(\ell+\ell^*) \le \ell$" both in terms of oscillation and the distinguished eigenvalue associated with ℓ. We turn now to finding a selfadjoint majorant for ℓ. According to Corollary 3.4 (with appropriate change of notation) we have the following.

4.10 Theorem. *Let $g_2(x)$ be sufficiently large to assure that the quadratic form*

$$Q_2(\xi) \equiv \sum_{i,j=1}^{n} P_{ij}\xi_i\xi_j + \xi_{n+1}\sum_{i=1}^{n} Q_i\xi_i + g_2\xi_{n+1}^2$$

is positive definite in D. If $u(x)$ is a solution $(L + g_2)u = 0$ which is positive in D and zero on ∂D, then every solution of $\ell v = 0$ has a zero in $\bar{D}$.

Thus in the sense of oscillatory behavior we have "$\ell \le \frac{1}{2}(\ell+\ell^*) + g_2$" with g_2 as defined above. We show next that this ordering is also correct in terms of distinguished eigenvalues. We let $\tilde{\Lambda}_1$ denote the first eigenvalue of $(L+g_2)u = \tilde{\Lambda}u$ in D, $u = 0$ on ∂D and let $u_1(x)$ denote the corresponding eigenfunction.

4.11 Theorem. *With λ_1 and $\tilde{\Lambda}_1$ defined as above $\lambda_1 \le \tilde{\Lambda}_1$.*

Proof. Suppose to the contrary that $\lambda_1 > \tilde{\Lambda}_1$. Then there exists $\tilde{\lambda}1$ such that $\lambda_1 > \tilde{\lambda}_1 > \tilde{\Lambda}_1$ and such that $\ell v = \tilde{\lambda}_1 v$ has a solution $v_1(x)$ which is positive in $\bar{D}$. Applying (3.3) with $p_{ij} \equiv P_{ij}$ and making use of the definition of ℓ and L yields

$$\sum_j \frac{\partial}{\partial x_j}\left[u_1 \sum_i p_{ij}\frac{\partial u_1}{\partial x_i} - \frac{u_1^2}{v_1}\sum_i p_{ij}\frac{\partial v_1}{\partial x_i}\right]$$

$$= u_1^2(\tilde{\lambda}_1 - \tilde{\Lambda}_1) + \left(g_2 - \sum_i \frac{\partial q_i}{\partial x_i}\right)u_1^2 - \frac{u_1^2}{v_1}\sum_i q_i \frac{\partial v}{\partial x_i}$$

$$+ \Sigma p_{ij}\left(\frac{\partial u_1}{\partial x_i} - \frac{u_1}{v_1}\frac{\partial v_1}{\partial x_i}\right)\left(\frac{\partial u_1}{\partial x_j} - \frac{u_1}{v_1}\frac{\partial v_1}{\partial x_j}\right).$$

Integrating over D and applying Green's theorem yields

$$0 = \iint_D (\widetilde{\lambda}_1 - \widetilde{\Lambda}_1) u_1^2 \, dx$$

$$+ \iint_D \left[p_{ij} \left(\frac{\partial u_1}{\partial x_i} - \frac{u_1}{v_1} \frac{\partial v_1}{\partial x_i} \right) \left(\frac{\partial u_1}{\partial x_j} - \frac{u_1}{v_1} \frac{\partial v_1}{\partial x_j} \right) + u_1 \sum_i q_i \left(\frac{\partial u_1}{\partial x_i} - \frac{u_1}{v_1} \frac{\partial v_1}{\partial x_i} \right) + g_2 u_1^2 \right] dx.$$

By the definition of $g_2(x)$, the second integral above is nonnegative, and we have the desired contradiction.

A somewhat different proof of Theorem 4.11 was given by Allegretto [2].

5. Maximum Principles

The maximum principle for selfadjoint elliptic equations deals with solutions of

$$(4.14) \qquad \ell u \equiv - \sum_{i,j=1}^{n} \frac{\partial}{\partial x_j} \left(p_{ij}(x) \frac{\partial u}{\partial x_i} \right) + p_o(x) u = 0$$

where ℓ is elliptic in a domain D. In its weak form it asserts if $p_o(x) \geq 0$ in D then a solution $u(x)$ attains its positive maximum on ∂D; in its strong form it asserts that if $p_o(x) \geq 0$ in D and $u(x)$ attains its positive maximum interior to D, then $u(x) \equiv$ constant in D.

These properties are intimately related to comparison theorems such as those discussed in Chapters 3 and 5. To illustrate this fact we shall show how to use Theorem 3.1 to establish a strong maximum principle.

Let $u(x)$ be a solution of (4.14) in a domain D and suppose it has a positive maximum at some $x_o \in D$. Then there exists a neighborhood D_o of x_o, $\overline{D}_o \subset D$, such that $u(x)$ is positive in $\overline{D}_o$ and satisfies $\frac{\partial u}{\partial \nu} + \sigma(x) u = 0$ on ∂D_o with $\sigma(x) \geq 0$ on ∂D_o. Consider now the comparison equation

$$(4.15) \qquad Lv \equiv - \sum_{i,j=1}^{n} \frac{\partial}{\partial x_j} \left(p_{ij}(x) \frac{\partial v}{\partial x_i} \right) = 0.$$

An obvious solution of (4.15) is $v(x) \equiv 1$, and this solution satisfies $\frac{\partial v}{\partial \nu} = 0$ on ∂D_o. Applying Theorem 3.1 we conclude that $u(x)$ is a constant multiple of $v(x)$ in

D_o. Assuming (4.14) is sufficiently regular so that its solutions have a unique continuation to D, we conclude that $u(x) \equiv u(x_o)$ in D.

This basic argument allows several generalizations. If the Sturmian comparison theorem is formulated for elliptic inequalities such as was done by Swanson [1], this argument leads to maximum principles for differential inequalities. In conjunction with a variation of Theorem 5.4 it leads to maximum principles for nonselfadjoint elliptic equations. Finally, the argument can also be applied to strongly elliptic systems, such as those to be discussed in Chapter 7, and used to formulate a maximum principle in this setting (see Kreith [13]).

In the other direction, Protter and Weinberger [1] have shown that maximum principles can be used to compare eigenvalues of nonselfadjoint elliptic equations of the form (3.5) and (3.6). Combined with the theory of positive operators to be developed in Chapter 5, these inequalities for eigenvalues make it possible to use maximum principles to establish Sturmian comparison theorems.

CHAPTER 5

ABSTRACT OSCILLATION THEORY

1. Positive Operators

In order to develop abstract formulations of the oscillation theory studied in the preceding chapters it is necessary to develop our abstract means of distinguishing between "positive solutions" and "solutions with a zero". Our means of abstracting these concepts to Banach spaces is to consider a cone of positive elements.

We consider a Banach space $\mathfrak{B}$; a closed subset $\mathfrak{P}$ of $\mathfrak{B}$ is called a <u>reproducing cone</u> if it satisfies

(i) $u \in \mathfrak{P}$ and $v \in \mathfrak{P} \Rightarrow u + v \in \mathfrak{P}$

(ii) $u \in \mathfrak{P}$ and $\alpha \geq 0 \Rightarrow \alpha u \in \mathfrak{P}$

(iii) $u \in \mathfrak{P}$ and $-u \in \mathfrak{P} \Rightarrow u = 0$

(iv) $w \in \mathfrak{B} \Rightarrow w = u - v$ for some $u,v \in \mathfrak{P}$.

Elements of $\mathfrak{P}$ are called <u>positive</u>, and we write $u \geq v$ if $u - v \in \mathfrak{P}$. A linear operator K mapping $\mathfrak{B}$ into $\mathfrak{B}$ is called <u>positive</u> if K maps $\mathfrak{P}$ into $\mathfrak{P}$.

The principal tool for establishing abstract comparison theorems in this setting is a generalization of the theorem of Perron and Frobenius on positive matrices to this more general setting. A positive operator A will be said to have property (*) if A has exactly one (normalized) eigenvector in $\mathfrak{P}$ corresponding to a simple positive eigenvalue which is larger than the absolute value of any other eigenvalue. The theorem of Perron and Frobenius asserts that if $\mathfrak{B}$ is E^n and A is represented by a matrix with positive elements, then A has property (*) with respect to the cone $\mathfrak{P}$ of vectors in E^n with nonnegative components. The simplest Banach space generalization deals with a cone $\mathfrak{P}$ having an interior $\mathfrak{P}^o$ and a compact positive operator A which maps $\mathfrak{P}$ into $\mathfrak{P}^o$; while such operators also possess property (*), the cones which will be of interest to us (e.g. the nonnegative functions square integrable on a domain D) will not have an interior. We shall therefore be in need of a more elaborate generalization based on the concept of u_o-positivity.

A bounded linear operator A is said to be u_o-positive with respect to $\mathcal{P}$ if there exists $u_o \in \mathcal{P}$, $u_o \neq 0$, such that for every $u \in \mathcal{P}$ there exists a constant $\delta > 0$ such that $\delta^{-1}u_o \leq Au \leq \delta u_o$. It is shown in Krasnoselskii [1; Chapter 2], that the following theorem is valid.

5.1 Theorem. *If A is compact and u_o-positive, then A has property (*). If v_o is the normalized positive eigenvector prescribed by (*), then A is v_o-positive.*

These results will enable an abstract formulation of Sturmian comparison theorems which yield new results for the special case of elliptic equations (see Kreith [8]).

2. An Abstract Comparison Theorem

We seek an abstract version of the Sturm comparison theorem for solutions of $\ell u = 0$ and $Lv = 0$, where ℓ and L are linear operators on a Banach space $\mathcal{B}$. Motivated by the discussion in Section 4, Chapter 4, we shall seek to formulate the hypotheses for such a theorem in terms of the resolvents of ℓ and L. The following preliminary result will be required.

5.2 Lemma. *Let A be a compact u_o-positive operator and μ and v_o be the maximal eigenvalue and corresponding normalized positive eigenfunction prescribed by (*). If there exists a $u \in \mathcal{P}$ $(u \neq \theta)$ such that $Au - \lambda u \in \mathcal{P}$ then $\lambda \leq \mu$. If $\lambda = \mu$, then u is a scalar multiple of v_o.*

Proof. Since $\mu > 0$, it is sufficient to deal with the case $\lambda > 0$. Recalling that $\mathcal{P}$ is closed by definition, we define

$$\epsilon_o = \sup\{\epsilon \mid A(v_o - \epsilon u) \in \mathcal{P}\}.$$

From $u \in \mathcal{P}$ $(u \neq \theta)$ and the v_o-positivity of A it follows that $0 < \epsilon_o < \infty$.

Writing

$$A(v_o-\epsilon_o u) = \mu v_o - \epsilon_o Au$$

$$= \mu v_o - \epsilon_o[\lambda u + (Au-\lambda u)]$$

we get

$$A[A(v_o-\epsilon_o u) + \epsilon_o(Au-\lambda u)] = \mu A(v_o - \epsilon_o \frac{\lambda}{\mu} u). \tag{5.2}$$

By hypothesis the left side of (5.2) belongs to $\mathcal{P}$, implying that $A(v_o - \epsilon_o \frac{\lambda}{\mu} u) \in \mathcal{P}$. From the definition of ϵ_o it follows that $\lambda \leq \mu$.

In case $\lambda = \mu$, (5.2) yields

$$\frac{1}{\mu} A^2(v_o-\epsilon_o u) \leq A(v_o-\epsilon_o u).$$

If $v_o - \epsilon_o u \neq \theta$, the v_o-positivity of A implies the existence of a $\delta > 0$ such that

$$\frac{1}{\mu} \delta^2 v_o \leq \frac{1}{\mu} A^2(v_o-\epsilon_o u) \leq A(v_o-\epsilon_o u) = \mu v_o - \epsilon_o Au.$$

But by the v_o-positivity of A, this contradicts the maximal property of ϵ_o, and therefore $v_o = \epsilon_o u$ whenever $\lambda = \mu$.

<u>5.3 Comparison Theorem</u>. <u>Let ℓ and L be linear operators on $\mathcal{B}$ and suppose there exists a constant $K > 0$ such that for $k \geq K$</u>

(i) $(\ell+kI)^{-1}$ <u>and</u> $(L+kI)^{-1}$ <u>are compact operators defined on all of $\mathcal{B}$;</u>

(ii) $(L+kI)^{-1} - (\ell+kI)^{-1}$ <u>is a positive operator;</u>

(iii) $(L+kI)^{-1}$ <u>is v_o-positive</u>.

<u>If $\ell u = 0$ has a nontrivial solution $u_o \in \mathcal{P}$ and if $Lu_o \neq 0$, then no nontrivial solution of $Lv = 0$ belongs to $\mathcal{P}$.</u>

<u>Proof</u>. Defining $A = (\ell+kI)^{-1}$ and $B = (L+kI)^{-1}$ we have $Au_o = \frac{1}{k} u_o$ for some nonzero $u_o \in \mathcal{P}$. Writing

$$(Bu_o - \frac{1}{k} u_o) = (B-A)u_o + (Au_o - \frac{1}{k} u_o)$$

we see by the preceding lemma that the maximal eigenvalue of B satisfies $\mu \geq \frac{1}{k}$. Since $\mu = \frac{1}{k}$ only if $Lu_o = 0$, we have $\mu > \frac{1}{k}$. Therefore v is a solution of $Lv = 0$ if and only if it satisfies $Bv = \frac{1}{k} v$ with $\frac{1}{k} < \mu$. Since B has property (*), the unique eigenvector of B in $\mathcal{P}$ corresponds to some $\mu > \frac{1}{k}$, and it follows that no nontrivial solution of $Lv = 0$ belongs to $\mathcal{P}$.

The most natural applications of this theorem are to elliptic operators ℓ and L which are sufficiently regular so that their resolvents can be represented as integral operators of the form

$$(\ell+kI)^{-1}u = \iint_D g_k(x,\xi)u(\xi)\, d\xi$$

$$(L+kI)^{-1}v = \iint_D G_k(x,\xi)u(\xi)\, d\xi.$$

In [8] the author shows that in this case there exists $k > 0$ such that hypotheses (i) and (iii) of Theorem 5.3 are satisfied. Letting ℓ and L be defined by

$$\ell u \equiv -\Sigma \frac{\partial}{\partial x_j}\left(p_{ij} \frac{\partial u}{\partial x_i}\right) + \Sigma q_i \frac{\partial u}{\partial x_i} + p_o u \quad \text{in } D$$

$$u = 0 \quad \text{on } \partial D$$

and

$$Lv \equiv -\Sigma \frac{\partial}{\partial x_j}\left(p_{ij} \frac{\partial v}{\partial x_i}\right) + \Sigma q_i \frac{\partial v}{\partial x_i} + (p_o - h)v$$

$$\frac{\partial v}{\partial \nu} + \sigma(x)v = 0 \qquad \text{on } \partial D$$

for $h(x) \geq 0$ ($h(x) \not\equiv 0$) and $\sigma(x)$ arbitrary, an examination of the Green's functions for ℓ and L shows that hypothesis (ii) of Theorem 5.3 is satisfied. Thus as a corollary of Theorem 5.3 one obtains the following comparison theorem for non-selfadjoint elliptic equations.

5.4 Theorem. *Let $u(x)$ be a nontrivial solution of $\ell u = 0$ and D a nodal domain for u. If $h(x) \geq 0$ in D ($h(x) \not\equiv 0$) and $v(x)$ is a nontrivial solution of $\ell v - hv = 0$, then $v(x)$ changes sign in D.*

Using Theorem 5.4, many of the results for selfadjoint elliptic equations in Chapter 4 can be extended to the nonselfadjoint case.

3. An Abstract Oscillation Theorem.

In order to generalize the concept of "$\ell u = 0$ is oscillatory at $x = \infty$" to an abstract operator ℓ, it is useful to view oscillation in terms of nodal domains. Using the fact that an ordinary differential equation of the form $\ell u = 0$ is oscillatory on an interval $[\alpha,\infty)$ if and only if it has an infinite number of disjoint nodal domains, we shall generalize the notion of oscillatory behavior to the case where ℓ generates a semibounded selfadjoint operator on a Hilbert space $\mathfrak{H}$. This form of abstract oscillatory behavior will lead to a property of the spectrum of selfadjoint realizations of ℓ which is well known in case ℓ is an ordinary differential operator (Dunford and Schwartz [1]).

Suppose ℓ is a symmetric linear operator defined on the domain D_ℓ in $\mathfrak{H}$. It is also assumed that ℓ is semibounded and satisfies $(\ell u,u) \geq ||u||^2$ for all $u \in D_\ell$. We recall (see Riesz and Nagy [1]) that ℓ has a distinguished selfadjoint extension $\bar{\ell}$, called the Friedrichs extension, which is obtained as follows: complete D_ℓ under the norm $|||u|||^2 = (\ell u,u)$ to construct a Hilbert space $\mathfrak{m}$ with inner product $((\cdot,\cdot))$; define $D_{\bar{\ell}}$ to consist of those elements $v \in \mathfrak{H}$ for which there exists a sequence u_n in D_ℓ such that

$$\lim_{n \to \infty} ||v-u_n|| = 0 \quad \text{and} \quad \lim_{m,n \to \infty} |||u_m-u_n||| = 0.$$

Then there exists a unique selfadjoint extension $\bar{\ell}$ of ℓ for which $D_{\bar{\ell}} \subseteq \mathfrak{H}$ and $((u,v)) = (u,\ell v)$ for all $u \in \mathfrak{m}$ and $v \in D_{\bar{\ell}}$.

In order to generalize the notion of nodal domain to this abstract situation we consider a closed linear subspace $\mathfrak{H}_k$ of $\mathfrak{H}$ for which $D_\ell \cap \mathfrak{H}_k$ is dense in $\mathfrak{H}_k$ and let J_k denote the projection operator with range $\mathfrak{H}_k$. In $\mathfrak{H}_k$ one can then define a symmetric operator ℓ_k as follows:

(i) $u \in D_{\ell_k}$ if $u \in D_\ell$ and $J_k u = u$,

(ii) if $u \in D_{\ell_k}$, then $\ell_k u = J_k \ell u$,

and denote by $\overline{\ell}_k$ the Friedrichs extension of ℓ_k. We now say that $\mathfrak{H}_k$ is a <u>nodal domain</u> for the adjoint equation

$$\ell^* u = \lambda u \tag{5.3}$$

if $\ell u = J_k \ell u$ for all $u \in \mathfrak{H}_k \cap D_\ell$ and[1]

$$x \in D_{\overline{\ell}_k} \frac{(\overline{\ell}_k u, u)_k}{||u||_k^2} = \lambda,$$

where this infimum is achieved by an eigenfunction $u_k \in D_{\overline{\ell}_k}$ and clearly $\lambda \geq 1$. Finally, a solution u_o of (5.3) is said to be <u>oscillatory</u> if there exist an infinite number of orthogonal closed subspaces such that

$$\mathfrak{H} = \sum_{k=0}^{\infty} \oplus \mathfrak{H}_k$$

and for $k \geq 1$ each $\mathfrak{H}_k$ is a nodal domain for (5.3) and $P_k u_o$ is the required eigenvector satisfying $(\overline{\ell}_k P_k u_o, P_k u_o)_k = \lambda ||P_k u_o||_k^2$.

In case ℓ is an appropriate ordinary or elliptic differential operator, the symmetric operator ℓ is obtained by defining D_ℓ to be $C_o^\infty(D)$ for some domain D. The following connection between oscillatory behavior and the spectrum of $\overline{\ell}$ has been noted for ordinary differential operators by Dunford and Schwartz [1] and Ladas [1]; the abstract version below also allows for ℓ to be a selfadjoint elliptic operator as well.

5.5 <u>Theorem</u>. <u>If a solution of (5.3) is oscillatory, then the essential spectrum of $\overline{\ell}$ intersects $[-\infty, \lambda]$.</u>

<u>Proof</u>. We shall show that given any $\epsilon > 0$ and any positive integer N, then there

---

[1] The subscript k indicates a norm or inner product in $\mathfrak{H}_k$.

exists a set of N linearly independent vectors $v_1,\cdots,v_N$ in D_ℓ such that for any $v = c_1v_1 + \cdots + c_nv_n \neq 0$ we have $(\ell v,v) < (\lambda+\epsilon)||v||^2$. Then by an argument used by Friedrichs [1], it follows that the spectrum of $\overline{\ell}$ is not discrete in $[1,\mu]$, the lower bound for $\overline{\ell}$ being at least 1. To construct $v_1,\cdots,v_N$ let $u_1,\cdots,u_N$ denote normalized eigenfunctions of $\overline{\ell}_1,\cdots,\overline{\ell}_N$ corresponding to the eigenvalue λ. By the definition of $\overline{\ell}_k$, for every u_k there exists a sequence u_{kj} satisfying

(i) $\quad u_{kj} \in D_\ell$ for $k=1,\cdots,N;\ j=1,2,\cdots$

(ii) $\quad J_k u_{kj} = v_{k\ell}$ for $k=1,\cdots,N;\ j=1,2,\cdots$

(iii) $\quad \lim\limits_{\ell\to\infty} |||P_k v_{kj} - u_k|||_k = 0$ for $k=1,\cdots,N$.

(iv) $\quad \lim\limits_{\ell\to\infty} ||P_k v_{kj} - u_k||_k = 0$ for $k=1,\cdots,N$.

We also have

$$|||v_{kj}||| = |||P_k v_{kj}|||_k \leq |||P_{kj} v_k - u_k|||_k + |||u_k|||_k$$

and

$$|||u_k|||_k = \lambda^{\frac{1}{2}} ||u_k||_k \leq \lambda^{\frac{1}{2}} (||u_k - P_k v_{kj}||_k + ||v_{kj}||).$$

Combining these yields

$$||v_{kj}|| \leq |||P_k v_{kj} - u_k|||_k + \lambda^{\frac{1}{2}} (||u_k - P_k v_{kj}||_k + ||v_{kj}||).$$

Using (iii) and (iv) and the fact that $\lim\limits_{j\to\infty} ||v_{kj}|| = 1$, one can choose j_o sufficiently large that

$$|||v_{kj_o}||| < (\lambda+\epsilon)^{\frac{1}{2}} ||v_{kj_o}|| \quad \text{for } k=1,\cdots,N.$$

Defining $v_k = v_{kj_o}$ and noting that

$$(v_j, v_k) = 0 \qquad \text{for } j \neq k$$

$$(\ell v_j, v_k) = 0 \qquad \text{for } j \neq k$$

$$(\ell v_k, v_k) < (\lambda+\epsilon)||v_k||^2$$

it follows that for any $v = c_1 v_1 + \cdots + c_n v_n \neq 0$

$$(\ell v, v) = \sum_{k=1}^{N} |c_k|^2 (\ell v_k, v_k) < (\lambda+\epsilon) \sum_{k=1}^{n} |c_k|^2 ||v_k||^2 = (\lambda+\epsilon)||v||^2$$

as was to be shown.

For second order ordinary differential operators Theorem 5.5 has a partial converse: if (5.3) has a nonoscillatory solution, then the essential spectrum of $\bar{\ell}$ does not intersect $(-\infty,\lambda)$. An example given by the author [9] for elliptic equations shows that such a converse cannot be expected to hold in general.

4. An Abstract Prüfer Transformation

The Prüfer transformation discussed in Chapter 1 has been the subject of recent study as the result of an interesting generalization introduced by Barrett [2]. In seeking to generalize the trigonometric functions $\sin \Theta(x)$ and $\cos \Theta(x)$ to matrix functions, Barrett considered the matrix system

$$\frac{dS}{dx} = Q(x)C \quad ; \quad \frac{dC}{dx} = -Q(x)S \tag{5.4}$$

where $S(x)$, $C(x)$ and $Q(x)$ are to be real $n \times n$ matrix functions, $Q(x)$ is symmetric, and $S(x)$, $C(x)$ satisfy the initial conditions

$$S(\alpha) = 0 \quad ; \quad C(\alpha) = I.$$

In the scalar case the solution of this system becomes

$$S(x) = \sin \int_{\alpha}^{x} Q(\xi)\, d\xi \quad ; \quad C(x) = \cos \int_{\alpha}^{x} Q(\xi)\, d\xi.$$

In general we shall denote the solution by $S(x) = S(x;\alpha,Q)$ and $C(x) = C(x;\alpha,Q)$, arguing existence by the standard Picard theory which is valid if $Q(x)$ is continuous and bounded on some interval $\mathcal{J}$.

The usefulness of these functions depends on a number of properties of $S(x;\alpha,Q)$ and $C(x;\alpha,Q)$ which were established by Barrett [2], Etgen [1], [2], and Reid [3].

<u>5.6 Theorem</u>. <u>For a given continuous symmetric matrix function</u> $Q(x)$ <u>the following identities hold on</u> $\mathcal{J}$:

(i) $C^*C + S^*S = I; \quad C^*S = S^*C$

(ii) $CC^* + SS^* = I; \quad CS^* = SC^*$.

<u>Proof</u>. The identities of (i) follow from the fact $\frac{d}{dx}(C^*C + S^*S) = \frac{d}{dx}(C^*S - S^*C) = 0$ and the initial conditions $S(\alpha) = 0$, $C(\alpha) = I$. To establish (ii) one defines the matrices

$$A = CC^* + SS^* \quad ; \quad B = CS^* - SC^*$$

and notes that

$$\frac{dA}{dx} = QB - BQ \quad ; \quad A(\alpha) = I$$

$$\frac{dB}{dx} = -QA + AQ \quad ; \quad B(\alpha) = 0$$

whose unique solution is $A \equiv I$, $B \equiv 0$.

From the identity $C^*S \equiv S^*C$ it follows readily that the "tangent function"

$$T(x;\alpha,Q) = S(x;\alpha,Q)C^{-1}(x;\alpha,Q)$$

is symmetric whenever $C(x;\alpha,Q)$ is nonsingular so that $T(x)$ exists. Viewing (5.4) as a special case of a Hamiltonian system, the symmetry of $T(x)$ reflects the fact that $(S(x),C(x))$ is a <u>conjoined</u> (or <u>prepared</u>) solution of (5.4). These ideas will be discussed in a more general context in Chapter 7.

Theorem 5.6 establishes the basic algebraic properties of $S(x)$ and $C(x)$ which will be needed to establish an abstract Prüfer transformation. However the question remains to what extent these generalized trigonometric functions exhibit oscillation properties of the scalar solutions $\sin \int_\alpha^x Q(\xi)\, d\xi$ and $\cos \int_\alpha^x Q(\xi)\, d\xi$.

For the case where $Q(x)$ is positive definite the following properties have been established by Etgen [1] and are stated here without proof:

1. If β, $\beta > \alpha$, is the first number such that $C(x;\alpha,Q)$ is singular, then $S(x;\alpha,Q)$ is nonsingular on (α,β).

2. If $\mathcal{J}$ is any closed interval on which $C(x)$ is nonsingular, then $S(x)$ has at most n singularities on $\mathcal{J}$. The roles of S and C may be interchanged.

3. If $\int_{\alpha}^{\beta} \mathrm{tr}\, Q < \pi/2$, then $S(x;\alpha,Q)$ is nonsingular on $(\alpha,\beta]$ and $C(x;\alpha,Q)$ is nonsingular or on $[\alpha,\beta]$.

4. If $\int_{\alpha}^{\beta} \mathrm{tr}\, Q < \pi$, then $S(x;\alpha,Q)$ is nonsingular on $(\alpha,\beta]$ and $C(x;\alpha,Q)$ has at most n singularities on $[\alpha,\beta]$.

5. A necessary and sufficient condition that $S(x;\alpha,Q)$ and $C(x;\alpha,Q)$ have at most a finite number of singularities on $[\alpha,\infty)$ is that $\int_{\alpha}^{\infty} \mathrm{tr}\, Q < \infty$.

These properties show that the Prüfer transformation to be developed below does indeed yield oscillation properties of the solution being represented.

While Barrett [1] studied matrix equations of the form $-(P_1(x)U')' + P_o(x)U = 0$, we shall follow Reid [2] and develop a Prüfer transformation for the more general system

$$\frac{dU}{dx} = G(x)W \quad ; \quad \frac{dW}{dx} = FU \tag{5.5}$$

where $G(x)$ and $F(x)$ are real symmetric $n\times n$ matrices whose components are assumed continuous and bounded on an appropriate interval $\mathcal{J} = [\alpha,\infty)$. The solution of (5.5) is to be a pair of $n\times n$ matrices $U(x)$, $W(x)$ satisfying the initial conditions $U(\alpha) = 0$, $W(\alpha) = W_o$, where W_o is nonsingular. Writing

$$U(x) = S^*(x)R(x) \quad ; \quad W(x) = C^*(x)R(x) \tag{5.6}$$

where $S(x)$ and $C(x)$ are the solutions of (5.4), one can readily apply Theorem 5.6 to verify that substitution into (5.5) yields

$$\frac{dR}{dx} = (SGC^* + CFS^*)R \quad ; \quad R(\alpha) = W_o \tag{5.7}$$

$$QR = (CGC^* - SFS^*)R. \tag{5.8}$$

If $Q(x)$ is a solution of (5.8) then it follows readily that a solution of (5.7) exists and is nonsingular on $\mathcal{J}$. However, since both $S(x) = S(x;\alpha,Q)$ and $C(x) = C(x;\alpha,Q)$, (5.8) is a functional equation for which the existence of a solution must be demonstrated. While such an existence proof is given by Barrett [2], Benson and Kreith [1] and Reid [4] circumvented this problem by redefining $S(x)$ and $C(x)$ by means of the system

$$\begin{aligned} \frac{dS}{dx} &= (CGC^* - SFS^*)C \quad ; \quad S(\alpha) = 0 \\ \frac{dC}{dx} &= -(CGC^* - SFS^*)S \quad ; \quad C(\alpha) = I \end{aligned} \tag{5.9}$$

which has a solution by the standard existence theory. The following result is then readily verified by differentiating (5.6).

5.7 Theorem. *If* $R(x)$ *and* $S(x)$, $C(x)$ *satisfy* (5.7) *and* (5.9), *respectively, then the Prüfer transformation* (5.6) *is valid on* $\mathcal{J}$.

The above method also has the advantage that it generalizes directly to the case where the functions involved take their values in a C*-algebra, while the original proof makes specific use of the matrix formulation of the problem (see Hille [1]). Another constructive proof of Theorem 5.7 is given by Reid [3] who also establishes that the matrices $Q(x)$ and $R(x)$ are determined by (5.5) up to unitary equivalence.

5. A Nontrigonometric Prüfer Transformation

The scalar Prüfer representation $u(x) = r(x) \sin \theta(x)$ raises the more general question of representing solutions of a Sturm-Liouville equation in terms of special functions other than the trigonometric functions. For example, is there a natural way of representing such solutions in the form $u(x) = r(x)J_n(\theta(x))$ where J_n is a Bessel function of order n? The existence of such a representation is topologically obvious. Letting $x_o < x_1 < x_2 < \cdots$ denote the (necessarily simple) zeros of $u(x)$, we need only choose a continuous strictly increasing function $\theta(x)$ such that $\theta(x_k) = j_{n,k}$, where $j_{n,k}$ is the k-th positive zero of J_n. The function $r(x)$ can then be determined in terms of $u(x)/J_n(\theta(x))$, with $r(x_k)$ being defined in an appropriate limiting sense.

The question being posed, however, is whether there is some way of normalizing this representation such that $r(x)$ and $\theta(x)$ are determined by means of equations analogous to (1.7) and (1.8). An affirmative answer to this question will be given in the context of the abstract theory of the preceding section. In place of the generalized sines and cosines defined by (5.4), we consider matrix valued functions $S(x)$, $C(x)$ and $s(x)$, $c(x)$ generated by the reciprocal systems

$$(5.10)\qquad \frac{dS}{dx} = Q(x)C \quad ; \quad \frac{dC}{dx} = -P(x)S$$
$$S(\alpha) = 0 \quad ; \quad C(\alpha) = I$$

and

$$(5.11)\qquad \frac{ds}{dx} = P(x)c \quad ; \quad \frac{dc}{dx} = -Q(x)s$$
$$s(\alpha) = 0 \quad ; \quad c(\alpha) = I,$$

where now $Q(x)$ and $P(x)$ are to be real, continuous, and symmetric matrix functions on $\mathcal{J}$. In order to achieve the representation $U(x) = S^*(x;\alpha,P,Q)R(x)$, $W(x) = c^*(x;\alpha,P,Q)R(x)$ we shall require some identities generalizing Theorem 5.6.

5.8 Theorem. *For given continuous symmetric matrix functions $Q(x)$ and $P(x)$, the following identities hold on $\mathcal{J}$:*

(i) $sS^* + Cc^* = I$

(ii) $sC^* - Cs^* = 0$

(iii) $cS^* - Sc^* = 0$

(iv) $S^*C - C^*S = 0$

(v) $s^*c - c^*s = 0$

(vi) $S^*s + C^*c = I$

(vii) $s^*S + c^*C = I$

Proof. To establish (i) - (iii) we define

$$A = sS^* + Cc^*$$

$$B = sC^* - Cs^*$$

$$D = cS^* - Sc^*$$

and note that A, B, and D satisfy the system

$$\frac{dA}{dx} = PD + BQ \quad ; \quad A(\alpha) = I$$

$$\frac{dB}{dx} = PA^* - AP \quad ; \quad B(\alpha) = 0$$

$$\frac{dD}{dx} = A^*Q - QA \quad ; \quad D(\alpha) = 0$$

whose unique solution is $A \equiv I$, $B = D \equiv 0$. The identities (iv) - (vii) are established in the same way as (i) of Theorem 5.6.

In order to determine functions $P(x)$, $Q(x)$ and $R(x)$ for which the representation $U = S^*R$, $W = c^*R$ is valid, we differentiate and use (5.5) to get

$$\frac{dU}{dx} = S^* \frac{dR}{dx} + C^*QR = Gc^*R$$

$$\frac{dW}{dx} = c^* \frac{dR}{dx} - s^*QR = FS^*R.$$

Using identities (i) - (iii) of Theorem (5.8) these equations can be solved for Q and R in the form

$$\frac{dR}{dx} = (sGc^* + CFS^*)R \tag{5.12}$$

$$QR = (cGc^* - SFS^*)R. \tag{5.13}$$

Since $R(\alpha) = W_o$ is nonsingular, $R(x)$ remains nonsingular on $\mathcal{J}$ and the second equation becomes

$$Q = cGc^* - SFS^*. \tag{5.13'}$$

It is readily verified by direct substitution that if $Q(x)$ and $R(x)$ satisfy (5.13') and (5.12), respectively, then the representation

$$U(x) = S^*(x)R(x) \quad ; \quad W(x) = c^*(x)R(x) \tag{5.14}$$

is valid _without any further conditions being imposed on_ $P(x)$. Various choices of $P(x)$ then make it possible to generate the desired special function $S(x)$ as a solution of (5.10).

As before there remains the problem of showing that the functional equation (5.13') has a solution. This is again accomplished by redefining $S(x)$, $C(x)$ and $s(x)$, $c(x)$ as solutions of

$$\frac{dS}{dx} = (cGc^* - SFS^*)C \quad ; \quad S(\alpha) = 0$$

$$\frac{dC}{dx} = -PS \quad ; \quad C(\alpha) = I$$

$$\frac{ds}{dx} = Pc \quad ; \quad s(\alpha) = 0$$

$$\frac{dc}{dx} = -(cGc^* - SFS^*)s \quad ; \quad c(\alpha) = I.$$

Making use of these relations, a direct calculation shows that the functions $U(x)$, $W(x)$ defined in (5.14) satisfy the differential system (5.5). These observations can be summarized as follows.

5.9 Theorem. _If_ $U(x)$, $W(x)$ _is a solution of_ (5.5), _then near_ $x = \alpha$ _this solution allows the representation_ (5.14), _where_ $Q(x)$ _and_ $R(x)$ _satisfy_ (5.13') _and_ (5.12), _respectively, and_ $P(x)$ _is an arbitrary continuous symmetric matrix function on_ $\mathcal{I}$.

As an application of this theorem we consider the problem of finding a Bessel function representation of a solution of the scalar system

$$\frac{du}{dx} = g(x)w \quad ; \quad \frac{dw}{dx} = f(x)u$$

(5.15)

$$u(\alpha) = 0 \quad ; \quad w(\alpha) = w_o \neq 0.$$

Choosing $Q(x) = \frac{1}{\theta}\frac{d\theta}{dx}$ and $P(x) = \frac{\theta^2-n^2}{\theta}\frac{d\theta}{dx}$, (5.10) and (5.11) become

$$\frac{dS}{dx} = \frac{1}{\theta}\frac{d\theta}{dx} C \quad ; \quad \frac{dC}{dx} = -\frac{\theta^2-n^2}{\theta}\frac{d\theta}{dx} S$$

(5.16)

$$S(\alpha) = 0 \quad ; \quad C(\alpha) = 1$$

and

$$\frac{ds}{dx} = \frac{\theta^2-n^2}{\theta}\frac{d\theta}{dx} c \quad ; \quad \frac{dc}{dx} - \frac{1}{\theta}\frac{d\theta}{dx} s$$

(5.17)

$$s(\alpha) = 0 \quad ; \quad c(\alpha) = 1,$$

respectively, so that $J_n(\theta(x))$ is the first component of the solution of (5.16). According to Theorem 5.9 we can achieve the representation $u(x) = r(x)J_n(\theta(x))$ by choosing $\theta(x)$ and $r(x)$ as solutions of

$$\frac{d\theta}{dx} = (g(x)c^2 - f(x)S^2)\theta$$

(5.18)

$$\theta(\alpha) = j_{n,1}$$

and

$$\frac{dr}{dx} = (g(x)sc + f(x)SC)r$$

(5.19)

$$r(\alpha) = w_o ,$$

respectively, where $j_{n,1}$ is the first positive zero of $J_n(x)$. As in the classical case, the equations (5.18) and (5.19) yield information regarding the oscillation properties and asymptotic behavior of $u(x)$ relative to $J_n(x)$.

CHAPTER 6

COMPLEX OSCILLATION THEORY

1. A Physical Interpretation

Let $p(x) = p_1(x) + ip_2(x)$ be a complex valued function of a real variable, continuous on an appropriate interval $\mathcal{I}$. The complex differential equation

$$-\frac{d^2u}{dx^2} + p(x)u = 0 \tag{6.1}$$

can then be decomposed into real and imaginary parts by writing $u(x) = u_1(x) + iu_2(x)$. The resulting system

$$\begin{aligned} -\frac{d^2u_1}{dx^2} + p_1(x)u_1 - p_2(x)u_2 &= 0 \\ -\frac{d^2u_2}{dx^2} + p_2(x)u_1 + p_1(x)u_2 &= 0 \end{aligned} \tag{6.2}$$

may be interpreted in terms of a particle of unit mass moving in the (u_1,u_2)-plane subject to a horizontal force $p_1(x)u_1 - p_2(x)u_2$ and a vertical force $p_2(x)u_1 + p_1(x)u_2$.

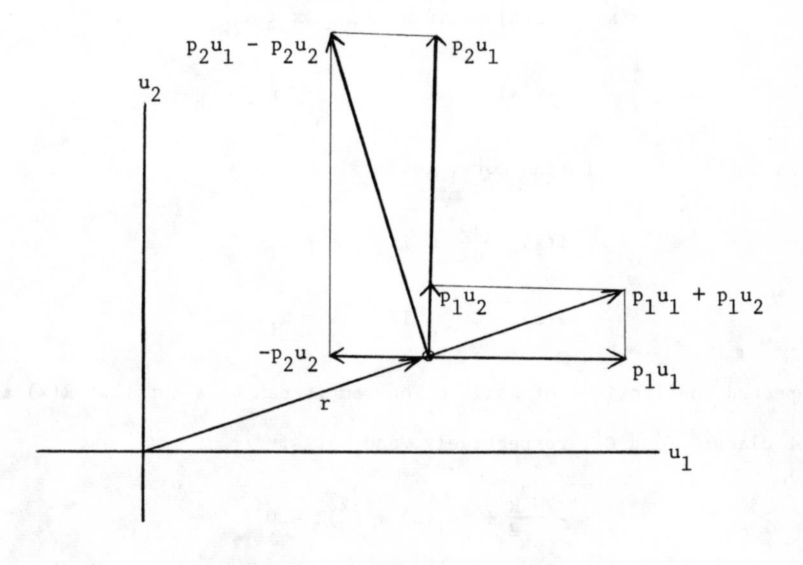

Figure 1 ($p_1 > 0$ and $p_2 > 0$)

Defining $r(x) = \sqrt{u_1^2(x) + u_2^2(x)}$, these forces can be resolved into a central force $p_1(x)r(x)$ away from the origin and a rotary force $p_2(x)r(x)$ perpendicular to the radius vector in a counterclockwise direction.

Such a physical interpretation can also be obtained by writing $u(x) = r(x)e^{i\theta(x)}$ and separating (6.1) into real and imaginary parts to obtain

$$(6.3) \qquad -\frac{d^2r}{dx^2} + \left[p_1(x) + \left(\frac{d\theta}{dx}\right)^2\right]r = 0$$

$$(6.4) \qquad -\frac{d}{dx}\left(r^2 \frac{d\theta}{dx}\right) + p_2(x)r^2 = 0.$$

These are just the radial and angular equations of motion of a particle subject to the forces described above. While this form provides a more convenient means for studying the behavior of solutions of (6.1), there is the difficulty that (6.3), (6.4) become singular whenever $u(x) = 0$. This difficulty was overcome by Taam [1] who defined an <u>associated solution</u> $U(x)$ of (6.1) as follows: let $\{\alpha_k\}$ denote the zeros of $u(x)$ on $\mathcal{I}$ and define

$$R(x) = r(x) \quad \text{if} \quad \alpha_{2k} \leq x \leq \alpha_{2k+1}$$

$$= -r(x) \quad \text{if} \quad \alpha_{2k-1} \leq x \leq \alpha_{sk} ,$$

where k is any integer. Furthermore, let

$$\Phi(x) = \frac{d\theta}{dx} \quad \text{if} \quad x \neq \alpha_i$$

$$\Phi(x) = 0 \quad \text{if} \quad x = \alpha_i .$$

Then by repeated applications of Rolle's Theorem it can be shown that $R(x)$ and $\Phi(x)$ are of class C^2 and C^1, respectively, and satisfy

$$(6.3') \qquad -\frac{d^2R}{dx^2} + (p_1(x) + \Phi^2)R = 0$$

$$-\frac{d}{dx}(R^2\Phi) + p_2(x)R^2 = 0 \tag{6.4'}$$

for all $x \in \mathcal{J}$.

The mechanical model discussed above suggests that oscillations of (6.1), defined in terms of zeros of $u(x)$, are an unusual phenomenon. In the real case, represented by a particle moving with only one degree of freedom, a sufficiently strong central force assures the existence of zeros for solutions of (6.1); however in the complex case we are asking that the solutions $u_1(x)$ and $u_2(x)$ have simultaneous zeros, and this is clearly more difficult to achieve. A simple example which illustrates the difficulties introduced by allowing complex valued solutions is the real equation $u'' + u = 0$. The solutions $\sin x$ and e^{ix} clearly violate the comparison and separation theorems established in Chapter 1.

These difficulties will reappear in a different form in Chapter 7 where higher order equations will be studied in terms of matrix differential equations of the form

$$\frac{dV}{dx} = AV + BZ$$

$$\frac{dZ}{dx} = CV + DZ.$$

Even in the case where all the matrices involved have real elements, the solutions will not in general be symmetric nor will they commute. This difficulty can be overcome in the case of selfadjoint problems, but it constitutes a major barrier to extending the oscillation theorems of Chapter 2 to nonselfadjoint equations of higher order.

2. Nonoscillation Theorems

While oscillatory behavior is difficult to establish for solutions of (6.1), there exists a variety of techniques for establishing nonoscillation. The following theorems due to Taam [1] are based on the equations (6.3') and (6.4') satisfied by the associated solutions $R(x)$ and $\Phi(x)$.

6.1 Theorem. *If $v(x)$ is a real solution of $-v'' + p_1(x)v = 0$ satisfying $v(x) \neq 0$ for $\alpha < x < \beta$ and $u(x)$ is a nontrivial solution of (6.1) with $p_2(x) \neq 0$, then $u(x)$ has at most one zero in $[\alpha,\beta]$.*

Proof. If $u(\gamma) = u(\delta) = 0$ for $\alpha \leq \gamma < \delta \leq \beta$, then by (6.4')

$$R^2(x)\Phi(x) = \int_\gamma^x p_2(\xi)R^2(\xi)\, d\xi \neq 0.$$

Applying the Sturm comparison theorem to

$$-\frac{d^2v}{dx^2} + p_1 v = 0$$

$$-\frac{d^2R}{dx^2} + (p_1+\Phi^2)R = 0$$

yields the desired result.

6.2 Theorem. *If $p_2(x)$ is of constant sign in $[\alpha,\beta]$ and $u(x)$ is a nontrivial solution of (6.1), then $u\frac{du}{dx}$ has at most one zero in $[\alpha,\beta]$.*

Proof. The zeros of $u(x)$ correspond to zeros of $R(x)$ while the zeros of $\frac{du}{dx}$ correspond to zeros of $\Phi(x)$. Therefore if $u\frac{du}{dx}$ is zero at γ and δ, $\alpha \leq \gamma < \delta \leq \beta$, then by (6.4')

$$0 = R^2\Phi\Big]_{x=\gamma}^{x=\delta} = \int_\gamma^\delta p_2(x)R^2(x)\, dx \neq 0$$

and we have the desired contradiction.

A somewhat different approach for establishing such nonoscillation theorems is based on the Green's transform as used by Hille [1]. It follows readily that

$$\frac{d}{dx}\left[\bar{u}\frac{du}{dx}\right] = \left|\frac{du}{dx}\right|^2 + p|u|^2$$

for any solution u(x) of (6.1). Separating into real and imaginary parts yields

$$\text{(6.5)} \qquad \operatorname{Re}\left[\bar{u}\,\frac{du}{dx}\right]_{\alpha}^{\beta} = \int_{\alpha}^{\beta}\left[\left|\frac{du}{dx}\right|^2 + p_1|u|^2\right] dx$$

$$\text{(6.6)} \qquad \operatorname{Im}\left[\bar{u}\,\frac{du}{dx}\right]_{\alpha}^{\beta} = \int_{\alpha}^{\beta} p_2|u|^2\, dx.$$

These equations lead to the following result.

6.3 Theorem. *If throughout* $\mathcal{J} = (\alpha,\beta)$ *either* $p_1(x) \geq 0$ *or* $p_2(x)$ *is of constant sign, then* $u\,\frac{du}{dx}$ *has at most one zero in* $\mathcal{J}$ *for any nontrivial solution* u(x) *of* (6.1).

The techniques of Taam and Hille both extend to the case of a complex independent variable and make it possible to establish zero free regions in the complex plane (see Taam [1], Hille [1], and Ince [1]). The Green's transform also has a natural generalization to matrix differential equations (see Chapter 7).

3. A Rotation Criterion

One means of trying to resolve the dilemma caused by the lack of oscillatory solutions is to study (6.1) in terms of a complex independent variable z

$$\text{(6.1')} \qquad -\frac{d^2u}{dz^2} + p(z)u = 0$$

and to seek zeros of solutions u(z) of (6.1') in the complex plane. While this leads to a more satisfactory analogue of the real theory, there does not appear to be any physical interpretation of such complex zeros. Another resolution can be sought in terms of the physical model described above and a different generalization of oscillatory behavior. Writing solutions of (6.1) in the polar form $u(x) = r(x)e^{i\theta(x)}$, we shall say that u(x) is *rotary at* $x = \infty$ if there exists an α such that $r(x) \neq 0$ for $x \in [\alpha,\infty)$ and $\lim_{t\to\infty} |\theta(t)| = \infty$. Regarding the origin as a critical point of the linear system (6.2) it is clear that (0,0) being a center or a stable focus would imply rotary behavior at $x = \infty$. However in general the concept of rotary behavior is independent of stability considerations. The following theorem establishes a link between rotary and oscillatory behavior.

6.3 Theorem. *If $p_2(x)$ is of constant sign for sufficiently large values of x and the real differential equation $-v'' + p_1(x)v = 0$ is oscillatory at $x = \infty$, then every nontrivial solution of (6.1) is rotary at $x = \infty$.*

Proof. By Theorem 6.2 every solution $u(x) = r(x)e^{i\theta(x)}$ satisfies $r(x) > 0$ for sufficiently large values of x. From (6.4) it follows that $\frac{d\theta}{dx}$ must also be of constant sign for sufficiently large values of x. Now let $v(x)$ be an oscillatory solution of $-v'' + p_1(x)v = 0$. Making use of (6.2) we write

$$-u_1v'' + p_1(x)u_1v = 0$$

$$-\ vu_1'' + p_1(x)u_1v - p_2(x)u_2v = 0$$

and subtract to obtain

$$\frac{d}{dx}(u_1v'-vu_1') - p_2(x)u_2v = 0.$$

If $\alpha < \beta$ are sufficiently large successive zeros of $v(x)$ and $v(x) > 0$ in (α,β), we integrate from α to β to obtain

$$u_1 \frac{dv}{dx} \Big|_\alpha^\beta = \int_\alpha^\beta p_2(x)u_2(x)v(x)\, dx. \tag{6.5}$$

To be specific, suppose $p_2(x) \geq 0$ (but not identically zero) for $\alpha \leq x \leq \beta$ and that $u_1(\alpha) \geq 0$, $u_2(\alpha) \geq 0$, corresponding to a particle in the first quadrant. If $u_1(x) \geq 0$, $u_2(x) \geq 0$ for $\alpha \leq x \leq \beta$, then the left side of (6.5) is negative whereas the right side is positive. This shows that if $(u_1(\alpha),u_2(\alpha))$ is in the first quadrant, then $(u_1(x),u_2(x))$ must rotate out of the first quadrant (in the direction dictated by the sign of $\frac{d\theta}{dx}$) during the interval (α,β). In case $p_2(x) \leq 0$, an analogous argument applies in the second quadrant. To complete the proof we observe that the system (6.3), (6.4) does not depend on θ. Therefore if $r(x)$, $\theta(x)$ is the solution of (6.3), (6.4) corresponding to the initial condition $r(\alpha) = r_o$, $r'(\alpha) = r_1$, $\theta(\alpha) = \theta_o$, $\theta'(\alpha) = \theta_1$, then $r(x)$, $\theta(x) - \gamma$ is the solution of (6.3) and (6.4) corresponding to the initial conditions $r(\alpha) = r_o$, $r'(\alpha) = r_1$, $\theta(\alpha) = \theta_o - \gamma$, $\theta'(\alpha) = \theta_1$. It follows that *every* nontrivial solution of (6.3), (6.4) must rotate in a fixed direction by at least $\pi/2$ radians while x increases from α to β. Since $v(x)$ is oscillatory, it follows that every nontrivial solution is rotary at ∞.

As an immediate consequence we have the following.

6.4 *Corollary*. *Under the hypotheses of Theorem* 6.3, *for every nontrivial solution* $u(x)$ *of* (6.1), $u_1(x)$ *and* $u_2(x)$ *are oscillatory at* $x = \infty$.

4. Complex Prüfer Transformations

In view of the insights afforded into real oscillation theory by the Prufer transformation, it is natural to seek extensions to the case of complex equations. The system

$$\frac{ds}{dz} = q(z)c; \qquad \frac{dc}{dz} = -q(z)s \tag{6.5}$$
$$s(\alpha) = 0 \quad ; \quad c(\alpha) = 1$$

has as in the real case the solution

$$s(z) = \sin \int_\alpha^z q(\xi)\, d\xi \quad ; \quad c(z) = \cos \int_\alpha^z q(\xi)\, d\xi .$$

The behavior of $s(z)$ can best be described by writing $q(z)$ in polar form, $q(z) = \rho(z)e^{i\varphi(z)}$ and defining a curve Γ in the complex plane by $z = F(t)$, where $F(t)$ is the solution of

$$\frac{dz}{dt} = e^{-i\varphi(z)} \quad ; \quad z(0) = \alpha. \tag{6.6}$$

Since $\left|\frac{dz}{dt}\right| \equiv 1$, t measures arc length on Γ and

$$s(F(t)) = \sin \int_{\alpha}^{t} \rho(\tau)\, d\tau.$$

Thus the zeros of $s(z)$ all lie on the curve Γ and occur whenever $\int_{\alpha}^{t} \rho(\tau)\, d\tau = n\pi$ for n any integer.

The identity $\sin^2 z + \cos^2 z \equiv 1$ allows one to achieve the usual Prufer representation

$$u(z) = r(z) \sin \theta(z) \quad ; \quad w(z) = r(z) \cos \theta(z)$$

for a complex system of the form

$$\begin{aligned} &\frac{du}{dz} = g(z)w \quad ; \quad \frac{dw}{dz} = f(z)u \\ &u(\alpha) = 0 \quad ; \quad w(\alpha) = w_o \,. \end{aligned} \tag{6.7}$$

That is, $\theta(z)$ and $r(z)$ are again determined by the differential equations (1.7) and (1.8), and $r(z) \neq 0$ for all z. Thus the zeros of $u(z)$ lie on the curve Γ which contains the zeros of $\sin \theta(z)$. Unfortunately, however, the differential equation

$$q(z) \equiv \frac{d\theta}{dz} = g(z) \cos^2 \theta(z) - f(z) \sin^2 \theta(z) \quad ; \quad \theta(\alpha) = 0$$

which determines $\theta(z)$ does not seem to provide readily any qualitative information which would help locate such a curve Γ or the location of zeros in the complex plane.

Another approach pursued by Barrett [3] consists of defining generalized trigonometric functions $s(x)$ and $c(x)$ on a real interval $\mathcal{J}$ by means of the equations

$$\frac{ds}{dx} = q(x)c \quad ; \quad \frac{dc}{dx} = -\bar{q}(x)s$$

(6.8)

$$s(\alpha) = 0 \quad ; \quad c(\alpha) = 1 \; .$$

From the fact that $\frac{d}{dx}[s\bar{s} + c\bar{c}] \equiv 0$, it follows that $|s|^2 + |c|^2 \equiv 1$ on $\mathcal{I}$ and one can again achieve a Prufer representation

(6.9) $$u(x) = r(x)s(x) \quad ; \quad w(x) = \bar{r}(x)\bar{c}(x)$$

for solutions of (6.7). The difficulty now is that the solutions of (6.8) are not well understood. Indeed Barrett shows that if $q(x) = \rho(x)e^{i\varphi(x)}$ and if $\varphi'(x)/\rho(x) = \text{constant} \neq 0$, then the solution $s(x)$ of (6.8) is oscillatory if and only if $\int^{\infty} |\rho(x)| \, dx = \infty$ whereas $c(x)$ has no zeros on $[\alpha,\infty)$!

It is of interest, however, that this Prüfer representation generalizes to a class of nonselfadjoint matrix differential equations which is not amenable to analysis by standard techniques (see Chapter 7 and Kreith [11]).

CHAPTER 7

DIFFERENTIAL EQUATIONS OF EVEN ORDER

1. Hamiltonian Systems

The oscillation theory discussed in Chapters 1 and 2 has a natural generalization to ordinary differential equations of even order. Our principal interest will be in selfadjoint equations of the form

$$Lv \equiv \sum_{k=0}^{n} (-1)^k (P_k(x) v^{(k)})^{(k)} = 0 \tag{7.1}$$

and the more general nonselfadjoint equation

$$\ell u \equiv \sum_{k=0}^{n} (-1)^k (p_k(x) u^{(k)})^{(k)} + \sum_{k=0}^{n-1} (-1)^k (q_k(x) u^{(k+1)})^{(k)} = 0 \tag{7.2}$$

where the coefficients $P_k(x)$, $p_k(x)$, $q_k(x)$ are for simplicity assumed to be of class C^k and $p_n(x)$, $P_n(x)$ to be positive on an appropriate interval $\mathcal{I}$.

There are numerous ways of representing an equation of the form (7.2) as a vector system. For the purpose of establishing oscillation properties it is most convenient to define

$$\begin{aligned} u_n &= p_n u^{(n)} \\ u_{n-k} &= -u'_{n-k+1} + q_{n-k} u^{(n-k+1)} + p_{n-k} u^{(n-k)}; \quad k=1,2,\cdots,n-1 \end{aligned}$$

so that (1.2) becomes

$$-u_1' + q_o u' + p_o u = 0. \tag{7.2'}$$

Defining column vectors

$$\begin{aligned} \underline{u}(x) &= \mathrm{col}(u(x), u'(x), \cdots, u^{(n-1)}(x)) \\ \underline{w}(x) &= \mathrm{col}(u_1(x), \cdots, u_n(x)). \end{aligned}$$

The equation (7.2) becomes

$$\begin{aligned} \underline{u}' &= a(x)\underline{u} + b(x)\underline{w} \\ \underline{w}' &= c(x)\underline{u} + d(x)\underline{w} \end{aligned} \tag{7.2''}$$

where $a(x)$, $b(x)$, $c(x)$ and $d(x)$ are $n \times n$ matrix functions defined by

$$a = (a_{ij}) \text{ where } a_{ij} = \begin{cases} 1 & \text{if } j=i+1 \\ 0 & \text{otherwise;} \end{cases}$$

$$b = \text{diag}(0,\dots, 0, 1/p_n);$$

$$c = (c_{ij}) \text{ where } c_{ij} = \begin{cases} p_{i-1} & \text{if } j=i, \\ q_{i-1} & \text{if } j=i+1, \\ 0 & \text{otherwise;} \end{cases}$$

$$d = (d_{ij}) \text{ where } d_{ij} = \begin{cases} -1 & \text{if } j=i-1, \\ q_{n-1}/p_n & \text{if } j=i=n, \\ 0 & \text{otherwise.} \end{cases}$$

Equation (7.1) allows an analogous representation

$$(7.1'') \qquad \begin{aligned} \underline{v}' &= A(x)\underline{v} + B(x)\underline{z} \\ \underline{z}' &= C(x)\underline{v} + D(x)\underline{z} \end{aligned}$$

in terms of vectors

$$\underline{v}(x) = \text{col}(v(x), v'(x),\dots, v^{(n-1)}(x))$$

$$\underline{z}(x) = \text{col}(v_1(x),\dots, v_n(x)),$$

with appropriate simplifications due to the selfadjointness of (7.1). In particular, in the selfadjoint case we have the conditions

$$A(x) \equiv -D^*(x), \quad B(x) \equiv B^*(x), \quad C(x) \equiv C^*(x)$$

which imply that (7.1") is a linear Hamiltonian system (see Coppel [1] for a more complete discussion of such systems).

In addition to vector systems of the form (7.1") and (7.2"), we shall have occasion to consider the matrix system

$$(7.3) \qquad \begin{aligned} V' &= A(x)V + B(x)Z \\ Z' &= C(x)V - A^*(x)Z \end{aligned}$$

associated with (7.1). Here $V(x)$ and $Z(x)$ are to be $n \times n$ matrix functions, so that V,

Z is a solution of (7.3) if and only if the corresponding columns of V, Z are solutions of (7.1").

An important property of (7.3) is that it admits a generalization of the concept of the Wronskian of two solutions.

<u>7.1 Lemma. If</u> V_1, Z_1 <u>and</u> V_2, Z_2 <u>are solutions of</u> (7.3), <u>then</u> $V_1^*Z_2 - Z_1^*V_2$ <u>is a constant matrix.</u>

<u>Proof</u>. The proof follows directly by differentiating $V_1^*Z_2 - Z_1^*V_2$ and substituting from (7.3).

While $V_1^*Z_2 - Z_1^*V_2$ is constant for any pair of solutions of (7.3), it is not necessarily the case that $V^*Z - Z^*V \equiv 0$ for a single solution. A solution V, Z which satisfies

$$V^*Z - Z^*V \equiv 0 \quad \text{for } x \in \mathcal{J}$$

will be called <u>conjoined on $\mathcal{J}$</u> (other terminology used includes <u>prepared</u> and <u>isotropic</u>).

The existence of prepared solutions of (7.3) is a consequence of the self-adjointness of (7.1) and makes it possible to generalize a variety of classical techniques to (7.3). These generalizations depend on the following simple observation.

<u>7.2 Lemma. If</u> V, Z <u>is a conjoined solution of</u> (7.3) <u>and</u> $V(x)$ <u>is nonsingular on</u> $\mathcal{J}$, <u>then</u> ZV^{-1} <u>is a symmetric matrix on</u> $\mathcal{J}$ -- i.e.

$$ZV^{-1} = V^{-1^*}Z^* \quad \text{for } x \in \mathcal{J}. \tag{7.4}$$

Finally we note the existence of a transformation which converts the Hamiltonian system (7.3) into the simpler system of the form

$$X' = G(x)Y; \quad Y' = F(x)X. \tag{7.5}$$

If $E(x)$ is a nonsingular solution of $E' = AE$, then the substitution $V = EX$, $Z = E^{*-1}Y$ into (7.3) yields (7.5) with $G = E^{-1}BE^{*-1}$ and $F = E^*CE$. A possible choice for E is given by

$$E = (E_{ij}) \text{ where } E_{ij} = \begin{cases} x^{j-1}/(j-i)! & \text{for } i \leq j \\ 0 & \text{for } i > j \end{cases}$$

so that E^{-1} is given by

$$E^{-1} = (E_{ij}^{-1}) \text{ where } E_{ij}^{-1} = \begin{cases} (-1)^{i+j} x^{j-i}/(j-i)! & \text{for } i \leq j \\ 0 & \text{for } i > j. \end{cases}$$

2. Comparison Theorems for Selfadjoint Equations

We first consider equations (7.1) and (7.2) in the special case where (7.2) is selfadjoint -- i.e. $q_k(x) \equiv 0$ for $k = 0,1,\cdots,n-1$. Given a real number α, the smallest $\beta > \alpha$ such that (7.1) has a nontrivial solution $v(x)$ satisfying

$$v(\alpha) = v'(\alpha) = \cdots = v^{(n-1)}(\alpha) = 0 = v(\beta) = v'(\beta) = \cdots = v^{(n-1)}(\beta)$$

is called the <u>first conjugate point of</u> α (with respect to L) and denoted by $\mu_1(\alpha)$. The conjugate point $\eta_1(\alpha)$ is analogously defined with respect to (7.2). If $\eta_1(\alpha) > \beta$, then (7.1) is said to be <u>disconjugate</u> on $[\alpha, \beta]$.

The relation between $\mu_1(\alpha)$ and the Hamiltonian system (7.3) is obtained by using solutions of (7.1) to construct a special prepared solution of (7.3) whose singularities correspond to the conjugate points of α. Let $v_1(x),\cdots,v_n(x)$ be a linearly independent set of solutions of (7.1) satisfying

$$v_i^{(j-1)}(\alpha) = 0 \quad ; \qquad i,j=1,\cdots,n,$$

$$v_{ij}(\alpha) = \delta_{ij} \quad ; \qquad i,j=1,\cdots,n.$$

Then every solution $v(x)$ of (7.1) satisfying

$$v(\alpha) = v'(\alpha) = \cdots = v^{(n-1)}(\alpha) = 0$$

is of the form $c_1v_1(x) + \cdots + c_nv_n(x)$. Furthermore the matrices V, Z whose ith columns are

$$\text{col}(v_i,\ v_i',\cdots,v_i^{(n-1)})$$

(7.6)

$$\text{col}(v_{i,1},\cdots,\ v_{i,n}),$$

respectively, are a conjoined solution of (7.3). Clearly the smallest $\beta > \alpha$ at which $c_1v_1(x) +\cdots+ c_nv_n(x)$ has an n-th order zero is the smallest $\beta > \alpha$ at which $V(x)$ becomes singular. In this way the conjugate points of α correspond to the singularities of a component of a conjoined solution of (7.3).

This observation leads to a direct generalization of the Sturm-Picone Theorem by means of a generalized Picone identity for (7.2") and (7.3). The application of this identity requires a generalized notion of matrix inversion for a symmetric nonnegative definite matrix B whose null space is orthogonal to the range of another matrix b. These conditions assure that the range of b is contained in the range of B and that there exists a matrix $K(x)$ such that $b = BK$. If $\underline{s}$ is in the range of b and $\underline{t}$ is such that $\underline{s} = b\underline{t}$, then we shall define $B^i b$ by $B^i b\underline{t} = K\underline{t}$. If B is nonsingular, then $B^i = B^{-1}$; B^i is in general not unique.

Given this generalized inverse we shall write $b \prec B$ in case

(i) The range of b is orthogonal to the null space of B, and

(ii) B has a generalized inverse B^i such that the symmetric part of $b^* - b^*B^ib$ is nonnegative definite.

7.3 Lemma. If V, Z is a conjoined solution of (7.3) with V nonsingular, $b \prec B$, and $\underline{u}$, $\underline{w}$ is a solution of (7.2") (with $q_k(x) \equiv 0$ for $k=0,1,\cdots,n-1$), then

$$\frac{d}{dx}[\underline{u}^*\underline{w} - \underline{u}^*ZV^{-1}\underline{u}] = \underline{u}^*(c - C)\underline{u} + \underline{w}^*(b^* - b^*B^ib)\underline{w} + \underline{y}^*B\underline{y},$$

where $\underline{y} = B^ib\underline{w} - ZV^{-1}\underline{u}$.

Proof. Expanding the left side of the identity to be established yields

$$\frac{d}{dx}[\underline{u}^*\underline{w} - \underline{u}^*ZV^{-1}\underline{u}] = \underline{u}^*\underline{w}' - \underline{u}^*Z'V^{-1}\underline{u} + \underline{u}^{*'}\underline{w} - \underline{u}^{*'}ZV^{-1}\underline{u} - \underline{u}^*ZV^{-1}\underline{u}' + \underline{u}^*ZV^{-1}V'V^{-1}\underline{u}.$$

Substituting from the differential equations (7.2') and (7.3), cancelling terms of the form $\underline{u}^*a\underline{w}$, $\underline{u}^*A^*ZV^{-1}\underline{u}$, $\underline{u}^*ZV^{-1}a\underline{u}$, inserting $-\underline{w}^*b^*B^{-1}b\underline{w} + \underline{w}^*b^*B^{-1}b\underline{w}$ and rearranging terms yields the desired identity. These manipulations require the symmetry of B and ZV^{-1}.

7.4 Theorem. Let $u(x)$ be a nontrivial solution of $\ell u \equiv \sum_{k=0}^{n}(-1)^k(p_k(x)u^{(k)})^{(k)} = 0$ satisfying $u(\alpha) = u'(\alpha) = \cdots = u^{(n-1)}(\alpha) = 0 = u(\beta) = \cdots = u^{(n-1)}(\beta)$. If

(i) $0 < P_n(x) \leq p_n(x)$ for $\alpha \leq x \leq \beta$

(ii) $P_k(x) \leq p_k(x)$ for $\alpha \leq x \leq \beta$ and $k=0,\cdots,n-1$,

and $\eta_1(\alpha)$ is the first conjugate point of α with respect to L, then $\eta_1(\alpha) \leq \beta$.

Proof. Suppose that $\eta_1(\alpha) > \beta$ so that the matrix V whose columns are given by (7.6) satisfies $\det V(x) \neq 0$ for $\alpha < x \leq \beta$. By continuity we may replace α by a slightly smaller $\alpha' < \alpha$ and thereby assure that $V(x) \neq 0$ for $\alpha \leq x \leq \beta$. Since $V(\alpha') = 0$, V, Z is conjoined and by Lemma 7.3

$$\frac{d}{dx}[\underline{u}^*\underline{w} - \underline{u}^*ZV^{-1}\underline{u}] \geq \underline{u}^*(c - C)\underline{u} + \underline{w}^*(b^* - b^*B^ib)\underline{w} \tag{7.7}$$

with equality if and only if $\underline{y}^*B\underline{y} = 0$. If $\underline{y}^*B\underline{y} = 0$, then

$$B^ib\underline{w} - ZV^{-1}\underline{u} = \underline{0}$$

which implies that

$$b\underline{w} - BZV^{-1}\underline{u} = \underline{0}$$

$$\underline{u}' - a\underline{u} - (V' - AV)V^{-1}\underline{u} = 0$$

$$\underline{u}' - V'V^{-1}\underline{u} = 0$$

$$(V^{-1}\underline{u})' = 0,$$

the last equation implying that $\underline{u} = V\underline{k}$ for some constant vector $\underline{k}$. But this contradicts the assumption that $\underline{u}(\alpha) = \underline{u}(\beta) = \underline{0}$ and that V is nonsingular on $[\alpha,\beta]$ and thereby rules out equality in (7.7). To show that inequality in (7.7) also yields a contradiction we integrate from α to β to obtain

$$0 > \int_\alpha^\beta [\underline{u}^*(c - C)\underline{u} + \underline{w}(b^* - b^*B^i b)\underline{w}]\, dx.$$

But hypotheses (i) and (ii) imply that $c - C$ and $b^* - b^*B^i b$ are nonnegative definite, and this completes the proof.

The generalized Picone identity of Lemma 7.3 lends itself to the generalizations and applications considered in Chapters 1 and 4. In particular, Theorem 4.1 has a direct generalization to selfadjoint differential equations of even order. The class of admissible functions now consists of functions $y(x)$ of class C^n satisfying

$$y(\alpha) = y'(\alpha) = \cdots = y^{(n-1)}(\alpha) = 0 = y(\beta) = \cdots = y^{(n-1)}(\beta).$$

7.5 Theorem. *If the differential equation* (7.1) *is disconjugate on* $[\alpha,\beta]$, *then*

$$\int_\alpha^\beta \sum_{k=1}^n p_k {y^{(k)}}^2\, dx > 0$$

for all admissible $y(x) \not\equiv 0$.

Another possible measure of oscillatory behavior is in terms of focal points which are defined as follows: The smallest $\beta > \alpha$ such that (7.1) has a nontrivial solution $v(x)$ satisfying

$$v(\alpha) = v'(\alpha) = \cdots = v^{(n-1)}(\alpha) = 0 = v_1(\beta) = \cdots = v_n(\beta)$$

is called the first (right) focal point of α with respect to L and denoted by $\eta_1(\alpha)$. Using analogous techniques, one can show that the hypotheses of Theorem 7.4 assure a comparison for focal points as well as conjugate points -- i.e. that $\eta_1(\alpha) \le \tilde{\mu}_1(\alpha)$. However, in the case of focal points one can sometimes weaken the hypotheses of Theorem 7.4 substantially, replacing the essentially pointwise inequalities among coefficients by integral conditions. In the case of second order equations such integral conditions have been established by Nehari [1] and Levin [1]. The generalization of Nehari's result to be sketched below is due to Travis [1] who considers the selfadjoint eigenvalue problems

$$(7.8) \qquad \ell u \equiv (-1)^n \left(p_n(x)u^{(n)}\right)^{(n)} = \lambda p_o(x)u$$

$$(7.9) \qquad Lv \equiv (-1)^n \left(P_n(x)v^{(n)}\right)^{(n)} = \Lambda P_o(x)v$$

subject to the boundary conditions

$$u(\alpha) = u'(\alpha) = \cdots = u^{(n-1)}(\alpha) = 0 = u_1(\beta) = \cdots = u_n(\beta)$$

$$v(\alpha) = v'(\alpha) = \cdots = v^{(n-1)}(\alpha) = 0 = v_1(\beta) = \cdots = v_n(\beta).$$

Due to the simplified form of the equations (7.8) and (7.9) we now have

$$u_k = \left(p_n u^{(n)}\right)^{(n-k)}; \quad v_k = \left(P_n v^{(n)}\right)^{(n-k)}.$$

<u>7.6 Theorem.</u> <u>If</u>

(i) $\quad 0 < p_o(x)$ on $[\alpha, \beta]$

(ii) $\quad \int_x^\beta p_o(\xi)\, d\xi \leq \int_x^\beta P_o(\xi)\, d\xi$ on $[\alpha, \beta]$

(iii) $\quad \int_\alpha^x \frac{1}{p_n(\xi)}\, d\xi \leq \int_\alpha^x \frac{1}{P_n(\xi)}\, d\xi$ on $[\alpha, \beta]$,

<u>then the first eigenvalues of</u> (7.8) <u>and</u> (7.9) <u>satisfy</u> $\Lambda_1 \leq \lambda_1$, <u>with equality if and only if</u> $p_n(x) \equiv P_n(x)$ <u>and</u> $p_o(x) \equiv P_o(x)$ <u>on</u> $[\alpha, \beta]$.

<u>Proof</u>. The proof makes use of the theory of positive operators presented in Chapter 5 and the fact that the Green's functions associated with the operators ℓ and L have the explicit representations

$$g(x,\xi) = \begin{cases} \dfrac{1}{(n-1)!}\displaystyle\int_\alpha^x \int_\alpha^{x_n} \cdots \int_\alpha^{x_2} \frac{(\xi-x_1)^{n-1}}{p_n(x_1)}\, dx_1 \cdots dx_n & \text{for } \alpha \leq x < \xi \leq \beta \\[2ex] \dfrac{1}{(n-1)!}\displaystyle\int_\alpha^\xi \int_\alpha^{x_n} \cdots \int_\alpha^{x_2} \frac{(x-x_1)^{n-1}}{p_n(x_1)}\, dx_1 \cdots dx_n & \text{for } \alpha \leq \xi < x \leq \beta \end{cases}$$

$$G(x,\xi) = \begin{cases} \dfrac{1}{(n-1)!}\displaystyle\int_\alpha^x \int_\alpha^{x_n} \cdots \int_\alpha^{x_2} \frac{(\xi-x_1)^{n-1}}{P_n(x_1)}\, dx_1 \cdots dx_n & \text{for } \alpha \leq x < \xi \leq \beta \\[2ex] \dfrac{1}{(n-1)!}\displaystyle\int_\alpha^x \int_\alpha^{x_n} \quad \int_\alpha^{x_2} \frac{(x-x_1)^{n-1}}{P_n(x_1)}\, dx_1 \cdots dx_n & \text{for } \alpha \leq \xi < x \leq \beta \end{cases}$$

A series of integration by parts now shows that the hypothesis (iii) assures that

$$0 \leq g(x,\xi) \leq G(x,\xi)$$

for $(x,\xi) \in [\alpha, \beta] \times [\alpha, \beta]$. We consider the Banach space

$$\mathcal{B} = \{u \in C^{2n}[\alpha, \beta] \mid u(\alpha) = u'(\alpha) = \cdots = u^{(n-1)}(\alpha) = 0\}$$

with norm

$$||u|| = \max_{\alpha \leq x \leq \beta} \{|u(x)|, |u'(x)|, \cdots, |u^{(2n)}(x)|\}$$

and the cones

$$\mathcal{P}_1 = \{u \in \mathcal{B} \mid u(x) \geq 0 \text{ on } [\alpha, \beta]\}$$

$$\mathcal{P}_2 = \{u \in \mathcal{P}_1 \mid u'(x) \geq 0 \text{ on } [\alpha, \beta]\}.$$

It can be shown (see Travis [1]) that the integral operator

$$M[u] = \int_\alpha^\beta g(x,\xi) p_o(\xi) u(\xi)\, d\xi$$

is u_o-positive in $\mathcal{P}_1$, where $u_o(x) = \int_\alpha^\beta g(x,\xi)p_o(\xi)\, d\xi$, and that if $p_o(x) \not\equiv 0$, $\int_x^\beta p_o(\xi)\, d\xi \geq 0$ on $[\alpha,\beta]$, then $M[u]$ is also u_o-positive with respect to $\mathcal{P}_2$. Defining

$$R[u] = \int_\alpha^\beta G(x,\xi)p_o(\xi)u(\xi)\, d\xi$$

it follows from the hypothesis that $M \leq R$ with respect to $\mathcal{P}_1$ while $R \leq N$ with respect to $\mathcal{P}_2$. Several applications of Lemma 5.2 now yield the desired result.

If in addition $P_o(x)$ is nonnegative, then the eigenvalues λ_1 and Λ_1 of (7.8) and (7.9) are monotone functions of the length of the interval $[\alpha,\beta]$ and we have the following.

7.7 Corollary. *If $\tilde{\mu}_1(\alpha)$ and $\tilde{\eta}_1(\alpha)$ are the first focal points of α with respect to ℓ and L, respectively, and if*

(i) $$\int_x^{\tilde{\mu}_1(\alpha)} p_o(\xi)\, d\xi \leq \int_x^{\tilde{\mu}_1(\alpha)} P_o(\xi)\, d\xi \qquad \text{on } [\alpha, \tilde{\mu}_1(\alpha)]$$

(ii) $$\int_\alpha^x \frac{1}{p_n(\xi)}\, d\xi \leq \int_\alpha^x \frac{1}{P_n(\xi)}\, d\xi \qquad \text{on } [\alpha, \tilde{\mu}_1(\alpha)],$$

then $\tilde{\eta}_1(\alpha) \leq \tilde{\mu}_1(\alpha)$.

3. Comparison Theorems for Nonselfadjoint Equations

Given the rather direct generalization of the Sturm-Picone Theorem to selfadjoint equations of the form (7.1), it is natural to ask whether this theory can be extended to a pair of nonselfadjoint equations of the form (7.2). We shall show below that it is relatively easy to compare a nonselfadjoint equation (7.2) with a selfadjoint equation (7.1) which oscillates slower and thereby to establish disconjugacy criteria for (7.2). However the other problem of establishing the existence of conjugate points for (7.2) by comparing it with (7.1) seems far less tractable, and it appears that criteria for the oscillation of equations of the form

(7.2) is one of the principal open questions in oscillation theory.

The difficulties encountered in dealing with nonselfadjoint equations result from the fact that if $q_k(x) \not\equiv 0$ for some k, then the system (7.2") is not self-adjoint in the sense that

$$b(x) \equiv b^*(x); \quad c(x) \equiv c^*(x); \quad a(x) \equiv -d^*(x).$$

Accordingly one cannot establish the existence of prepared solutions of the matrix system

$$\begin{aligned} U' &= a(x)U + b(x)W \\ W' &= c(x)U + d(x)W \end{aligned} \tag{7.10}$$

nor the symmetry of WU^{-1} whenever U is nonsingular. The condition $WU^{-1} = (WU^{-1})^*$ would be essential in using the techniques of Theorem 7.4 to establish the existence of conjugate points for (7.2). The lack of such symmetry is analogous to allowing complex valued solutions as was done in Chapter 6. As in the case of complex valued solutions, one can establish nonoscillation for systems such as (7.10) but not oscillation.

In establishing nonoscillation criteria for (7.2) one seeks conditions of the form "$L \leq \ell$" which will assure that $\eta_1(\alpha) \leq \mu_1(\alpha)$. Let $u(x)$ be a nontrivial solution of (7.2) satisfying

$$u(\alpha) = u'(\alpha) = \cdots = u^{(n-1)}(\alpha) = 0 = u(\beta) = \cdots = u^{(n-1)}(\beta).$$

The corresponding vector solution $\underline{u}(x)$, $\underline{w}(x)$ of (7.2") then satisfies $\underline{u}(\alpha) = \underline{u}(\beta) = \underline{0}$ and we again seek a generalized Picone identity to handle this situation. The following generalization of Lemma 7.3 is our principal tool.

<u>7.8 Lemma</u>. <u>Suppose</u> $\underline{u}$, $\underline{w}$ <u>is a nontrivial solution of</u> (7.2") <u>and</u> V, Z <u>is a conjoined solution of</u> (7.3) <u>for which</u> V <u>is nonsingular</u>. <u>If</u> $b < B$, <u>then</u>

$$\frac{d}{dx}[\underline{u}^*\underline{w} - \underline{u}^*ZV^{-1}\underline{u}] = \underline{u}^*(c-C)\underline{u} + \underline{u}^*[a^* + d + (a^*-A^*)B^i b]\underline{w} + \underline{w}^*(B^i b)^*(A-a)\underline{u}$$

$$+ \underline{w}^*(b^* - b^*B^i b)\underline{w} + \underline{u}^*(a^*-A^*)\underline{y} + \underline{y}^*(a-A)\underline{u} + \underline{y}^*B\underline{y} \; ,$$

where $\underline{y} = B^i b\underline{w} - ZV^{-1}\underline{u}$.

The proof is analogous to that of Lemma 7.3 with details given in Kreith [10]. In applying this identity to the special systems under consideration we have

$$A = a$$

$$a^* + d = \mathrm{diag}(0, \cdots, 0, q_{n-1}/p_n)$$

$$c - C = (\gamma_{ij}) \quad \text{where } \gamma_{ij} = \begin{cases} p_{i-1} - P_{i-1} & \text{if } j=i \\ q_{i-1} & \text{if } j=i+1 \\ 0 & \text{otherwise} \end{cases}$$

$$b^* - b^*B^i b = \mathrm{diag}\left(0, \cdots, 0, \frac{p_n - P_n}{p_n^2}\right).$$

In order to generalize the method of proof of Theorem 7.4, it is necessary to assure that the symmetric part of the 2n×2n matrix

$$\begin{pmatrix} c - C & a^* + d \\ 0 & b^* - b^*B^i b \end{pmatrix}$$

is nonnegative definite, and these conditions are precisely the hypothesis in the following.

7.9 Theorem. *If for* $\alpha \leq x \leq b$

(i) $p_n(x) \geq P_n(x) > 0$ *and* $p_n(x) > P_n(x)$ *on* $\{x | q_{n-1}(x) \neq 0\}$

(ii) $c - C - \mathrm{diag}\left(0, \cdots, 0, \frac{q_{n-1}^2}{4(p_n - P_n)}\right)$ *has a nonnegative definite symmetric part*

then $\eta_1(\alpha) \leq \beta$.

Another means of obtaining disconjugacy criteria for (7.2) is to use operator ordering techniques of Chapter 4 and Theorem 7.5. With ℓ given by (7.2) we find that $\ell + \ell^*$ is defined by

$$\frac{1}{2}(\ell+\ell^*)u = \sum_{k=0}^{n} (-1)^k (p_k(x)u^{(k)})^{(k)} + \sum_{k=0}^{n-1} (-1)^k [(p_k(x) - \tfrac{1}{2}q_k'(x))u^{(k)}]^{(k)}.$$

If u(x) is a nontrivial solution of (7.2) which realizes the conjugate point $\beta = \mu_1(\alpha)$, then integration by parts shows that

$$\mathcal{J}[u] \equiv \int_\alpha^\beta \left[p_n(u^{(n)})^2 + \sum_{k=0}^{n-1} (p_n - \tfrac{1}{2}q_k')(u^{(n)})^2\right] dx$$

$$= \int_\alpha^\beta u\ell u \, dx = 0.$$

Therefore

$$\int_\alpha^\beta \left[(p_n - P_n)(u^{(n)})^2 + \sum_{k=0}^{n-1} (p_k - \tfrac{1}{2}q_k' - P_k)(u^{(k)})^2\right] dx$$

(7.11)

$$+ \int_\alpha^\beta \sum_{k=1}^{n} P_k (u^{(k)})^2 \, dx = 0.$$

An application of Theorem 7.5 now yields the following.

7.10 Theorem. *If for* $\alpha \leq x \leq \beta$

(i) $p_k(x) \geq P_k(x) + \frac{1}{2}q_k'(x)$; $k=0,1,\cdots,n-1$

(ii) $p_n(x) \geq P_n(x) > 0$,

then $\eta_1(\alpha) \geq \beta$.

The apparent difference between Theorems 7.9 and 7.10 can be resolved by an appropriate modification of the device of Picard. If $h_o(x), h_1(x), \cdots, h_{n-1}(x)$

are continuously differentiable and $u(x)$ has $(n-1)$st order zeros at α and β, then

$$0 = \int_\alpha^\beta \frac{d}{dx}\left[h_k(u^{(k)})^2\right] dx = \int_\alpha^\beta \left[h_k'(u^{(k)})^2 + 2h_k u^{(k)} u^{(k+1)}\right] dx.$$

Adding this to (7.11) and choosing $h_k = \frac{1}{2}q_k$ yields

$$\int_\alpha^\beta \left[(p_n - P_n)(u^{(n)})^2 + \sum_{k=0}^{n-1}(p_k - P_k)(u^{(k)})^2 + \sum_{k=0}^{n-1} q_k u^{(k)} u^{(k+1)}\right] dx$$

$$+ \int_\alpha^\beta \sum_{k=0}^{n} P_k (u^{(k)})^2 \, dx = 0$$

which leads to Theorem 7.9.

While there do not appear to be any known oscillation criteria for (7.2) in terms of conjugate points, it is possible to show the existence of focal points for nonselfadjoint equations of even order. We consider two nonselfadjoint equations of the form (7.2) and

$$(7.12) \qquad Lv \equiv \sum_{k=0}^{n} (-1)^k (P_k(x) v^{(k)})^{(k)} + \sum_{k=0}^{n-1} (-1)^k (q_k(x) v^{(k+1)})^{(k)} = 0.$$

These two equations can be formulated as vector systems of the form (7.2") and (7.1"). Furthermore the focal points $\tilde{\mu}_1(\alpha)$ and $\tilde{\eta}_1(\alpha)$ of α with respect to ℓ and L, respectively, can be given by the first singularities of matrices $W(x)$ and $Z(x)$, where U, W and V, Z are appropriately chosen matrix solutions of the matrix system associated with (7.2") and (7.1"), respectively.

Consider now the transformations

$$M(x) = U(x)W^{-1}(x); \quad N(x) = V(x)Z^{-1}(x).$$

The matrices $M(x)$ and $N(x)$ satisfy the matrix Riccati equations

$$\frac{dM}{dx} = b(x) - Mc(x)M + a(x)M - Md(x); \; M(\alpha) = 0. \tag{7.13}$$

$$\frac{dN}{dx} = B(x) - NC(x)N + A(x)M - MD(x); \; N(\alpha) = 0. \tag{7.14}$$

In order to show that $\tilde{\eta}_1(\alpha) \le \tilde{\mu}_1(\alpha)$ it is sufficient to show that the solution M(x) of (7.13) can be continued at least as far as the solution N(x) of (7.14). Since M(x) and N(x) will not in general be symmetric, the usual arguments based on positive definiteness will not apply. One can, however, establish the following.

7.11 Theorem. *If "$\ell \ge L$" in the sense that for $\alpha \le x \le \tilde{\mu}_1(\alpha)$*

(i) $p_n(x) \ge P_n(x) > 0$

(ii) $0 \ge p_k(x) \ge P_k(x)$; $k=0,1,\cdots,n-1$

(iii) $0 \ge q_k(x) \ge Q_k(x)$; $k=0,1,\cdots,n-1$,

then $\tilde{\eta}_1(\alpha) \le \tilde{\mu}_1(\alpha)$.

Proof. Following the notation of Reid [5], we write $G \cdot\!\ge\!\cdot 0$ if G is a square $n \times n$ matrix with elements $G_{ij} \ge 0$ for $i,j = 1,\cdots,n$. Similarly $G \cdot\!\ge\!\cdot F$ denotes $G - F \cdot\!\ge\!\cdot 0$. Our hypotheses assure that

$$B(x) \cdot\!>\!\cdot b(x) \cdot\!\ge\!\cdot 0$$

$$-C(x) \cdot\!\ge\!\cdot -c(x) \cdot\!\ge\!\cdot 0$$

$$A(x) \cdot\!\ge\!\cdot a(x) \cdot\!\ge\!\cdot 0$$

$$-D(x) \cdot\!\ge\!\cdot -d(x) \cdot\!\ge\!\cdot 0$$

so that the solution of (7.13) and (7.14) satisfy $N(x) \cdot\!\ge\!\cdot M(x) \cdot\!\ge\!\cdot 0$. By the Perron-Frobenius theory of nonnegative matrices, this assures that the distinguished (positive) largest eigenvalue of N(x) is at least as large as the largest eigenvalue of M(x) and that interval of existence of N(x) is no larger than that of M(x).

This argument does not apply to conjugate points because the right sides of the equations corresponding to (7.13) and (7.14) are not nonnegative in the ordering induced by $\cdot\!\ge\!\cdot$.

4. A Prüfer Transformation for Nonselfadjoint Systems

The abstract Prüfer transformation introduced in Chapter 5 has a generalization to nonselfadjoint matrix (or B*) valued systems of the form (7.2) (see Kreith [11]). Motivated by the complex scalar Prüfer transformation of Barrett [3] we consider generalized sines and cosines $s(x)$, $C(x)$ and $S(x)$, $c(x)$ defined by the reciprocal systems

$$\frac{ds}{dx} = Q(x)C \quad ; \quad \frac{dC}{dx} = -Q^*(x)s$$

$$s(\alpha) = 0 \quad ; \quad C(\alpha) = I$$

(7.15)

$$\frac{dS}{dx} = Q^*(x)c \quad ; \quad \frac{dc}{dx} = -Q(x)S$$

$$S(\alpha) = 0 \quad ; \quad c(\alpha) = I ,$$

respectively, where $Q(x)$ is a continuous matrix function on $\mathcal{I}$. In case $Q(x)$ is nonsingular, $s(x)$ and $c(x)$ allow a simpler definition in terms of the differential equation

$$\frac{d}{dx}\left(Q^{-1}(x)\frac{dy}{dx}\right) + Q^*(x)y = 0.$$

The Prüfer representation

$$U(x) = s^*(x)R(x) \quad ; \quad W(x) = c^*(x)R(x) \tag{7.16}$$

follows in an analogous manner to that of (5.6), except that the identities of Theorem 5.6 must be replaced by the following

7.12 Theorem. *If* $s(x)$, $C(x)$ *and* $S(x)$, $c(x)$ *satisfy* (7.15) *on* $\mathcal{I}$, *then*

(i) $s^*s + C^*C \equiv I$

(ii) $C^*S - s^*c \equiv 0$

(iii) $ss^* + cc^* \equiv I$

(iv) $\quad SS^* + CC^* \equiv I$

(v) $\quad Sc^* - Cs^* \equiv 0$

(vi) $\quad sC^* - cS^* \equiv 0$

on $\mathcal{J}$.

Proof. Identities (i) and (ii) follow from the fact that

$$\frac{d}{dx}(s^*s+C^*C) = \frac{d}{dx}(C^*S-s^*c) = 0$$

on $\mathcal{J}$. To establish (iii) - (vi) we define $A = ss^* + cc^*$, $B = SS^* + CC^*$, $D = Sc^* - Cs^*$, and $E = sC^* - cS^*$ and note that A, B, D, and E satisfy the system

$$\frac{dA}{dx} = EQ^* - QD \quad ; \quad A(\alpha) = I$$

$$\frac{dB}{dx} = DQ - Q^*E \quad ; \quad B(\alpha) = I$$

$$\frac{dD}{dx} = Q^*A - BQ^* \quad ; \quad D(\alpha) = 0$$

$$\frac{dE}{dx} = QB - AQ \quad ; \quad E(\alpha) = 0$$

which has the obvious solution $A = B \equiv I$, $D = E \equiv 0$ on $\mathcal{J}$.

The remainder of the development parallels that of Chapter 5. Whether this representation is useful in studying the behavior of solutions of nonselfadjoint systems is at this point not clear. As demonstrated by Barrett [3], the behavior of these generalized trigonometric functions is not well understood even in the scalar case, and this fact restricts the usefulness of the representation (7.16).

5. Oscillation Criteria

A selfadjoint differential equation of the form (7.1) defined on $\mathcal{J} = [\alpha,\infty)$, is said to be oscillatory at ∞ if $\eta_1(\beta) < \infty$ for any $\beta \geq \alpha$. In view of the connection established between conjugate points with respect to L and the singularities of a conjoined solution of a related Hamiltonian system, the oscillatory behavior of (7.1) at $x = \infty$ can be studied in terms of systems of the form (7.3) or (7.5).

The singularities of conjoined solutions of (7.3) and (7.5) have been studied by means of matrix Riccati equations by Reid [2], Ahlbrandt [1] and Tomastik [1]. Reid has also made extensive use of a variational criterion for oscillation analogous to that of Theorem 4.1, and these results are surveyed in [6]. In this section we restrict our attention to an application of Lemma 7.3, generalizing slightly a result of the author [12]. Similar results have been obtained by Swanson [6].

Let u(t), w(t) be a nontrivial solution of the scalar system

$$\frac{du}{dx} = g(x)w \quad ; \quad \frac{dw}{dx} = f(x)u \tag{7.17}$$

on an interval $\mathcal{I} = [\alpha,\infty)$ and for which u(x) is oscillatory at $x = \infty$. If J is any constant n×n matrix, then the matrix functions $U(x) = u(x)J$, $W(x) = w(x)J$ satisfy the matrix system

$$\frac{dU}{dx} = g(x)IW \quad ; \quad \frac{dW}{dx} = f(x)IU \tag{7.17'}$$

as well as $U(x_1) = U(x_2) = \cdots = 0$ for a sequence $\{x_n\}$ where $x_n \uparrow \infty$. The proof of Lemma 7.3 readily applies to matrix solutions U, W as well as vector solutions $\underline{u}$, $\underline{w}$, yielding the following result.

<u>7.13 Lemma</u>. <u>If</u> V, Z <u>is a conjoined solution of</u>

$$\frac{dV}{dx} = G(x)Z \quad ; \quad \frac{dZ}{dx} = F(x)V \tag{7.18}$$

<u>with</u> V <u>nonsingular</u>, $g < G$, <u>and</u> U, W <u>is a solution of</u> (7.17'), <u>then</u>

$$\frac{d}{dx}[U^*W-U^*ZV^{-1}U] = uJ^*(fI-F)Ju$$
$$+ wJ^*(gI-gG^i g)Jw + Y^*GY,$$

<u>where</u> $Y = G^i gW - ZV^{-1}U$.

We recall that by Theorem 1.3 the conditions $g(x) > 0$ and

$$\int^{\infty} g(x)\, dx = -\int^{\infty} f(x)\, dx = \infty \tag{7.19}$$

assure the existence of a nontrivial solution u, w of (7.16) such that u(x) is oscillatory at $x = \infty$. Combining (7.19) with the choice of appropriate constant matrices J which assure that $J^*(fI-F)J$ and $J^*(gI-gG^{i}g)J$ are nonnegative definite leads to oscillation criteria for V(x). By means of the representation (7.5), these conditions can be translated into oscillation criteria for (7.1). This process is quite cumbersome and will not be pursued in detail here. We content ourselves to state without proof a result which one obtains by applying this technique to fourth order equations.

7.14 Theorem. *If* $n = 2$ *and there exist functions* $f(x)$ *and* $g(x) > 0$ *such that*

(i) $-\int^{\infty} f(x)\, dx = \int^{\infty} g(x)\, dx = \infty$

(ii) $\frac{1}{p_2(x)} \geq g(x)$

(iii) $x^2 p_o(x) + p_1(x) \leq f(x)$

on some interval $[\alpha,\infty)$, *then* (7.1) *is oscillatory at* $x = \infty$.

6. An Inverse Sturm-Liouville Problem

The inverse Sturm-Liouville problem to be considered here is that of determining the coefficients of a differential equation from knowledge regarding the oscillatory behavior of its solutions. In the case of the selfadjoint Sturm-Liouville equation

$$-(p_1(x)u')' + p_o(x)u = 0 \tag{7.20}$$

this problem has been solved by Martin [1] who considers a solution

$w(x) = w(x;c,\Theta)$ of (7.20) satisfying

(7.21) $$B(w,c,\Theta) \equiv w(c) \cos \Theta - p_1(c)w'(c) \sin \Theta = 0.$$

Martin shows that knowledge of the first zero of $w(x;c,\Theta)$ determines the coefficients of (7.20). Using the relation $w(f;c,\Theta) = 0$ to define implicitly a function $x = f(c,\Theta)$, explicit formulas can be derived for $p_1(x)$ and $p_o(x)$ in terms of $f(c,\Theta)$. In the special case $p_1(x) \equiv 1$, $p_o(x) < 0$ Fink [1] has shown more directly that $f(c,\Theta)$ determines $p_o(x)$ uniquely.

In order to allow a generalization to higher order equations, we begin by giving a somewhat different development of Martin's results. Solving (7.21) for Θ yields $\Theta = \arctan w/p_1w$, and this expression will be referred to as "the Θ-value of w at x". The function $x = f(c,\Theta)$ prescribes implicitly the Θ-value of w which is required at c to produce a zero at x. It is knowledge of $f(c,\Theta)$ for $c \in \mathcal{J}$ and Θ near zero which is required to determine $p_1(x)$ and $p_o(x)$ in $\mathcal{J}$.

Consider now the function $\Theta(x) = \arctan w/p_1w$. Given a solution $u(x;c)$ of (7.20) with a zero at $x = c$, $g(x,c) \equiv \arctan {}^{u(x;c)}\!/_{p_1(x)u'(x;c)}$ prescribes the Θ-value of u at x. Since $\lim_{\Theta \to 0} f(c,\Theta) = c$ and $\lim_{x \to c} g(x,c) = 0$, we have the following proposition:

<u>knowledge of</u> $f(c,\Theta)$ <u>in a neighborhood of</u> $\Theta = 0$ <u>is equivalent to knowledge of</u> $g(x,c)$ <u>in a neighborhood of</u> $x = c$. Our development differs from Martin's insofar as we derive formulas for $p_1(x)$ and $p_o(x)$ in terms of $g(x,c)$ rather than $f(c,\Theta)$.

<u>7.15 Theorem</u>. <u>If the function</u> $g(x,c)$ <u>is known for all</u> x <u>in a neighborhood of</u> c <u>and all</u> $c \in \mathcal{J}$, <u>then the coefficients of</u> (7.20) <u>are determined uniquely</u>.

<u>Proof</u>. If there were two sets of coefficients $p_{1,i}(x)$ and $p_{o,i}(x)$ $(i=1,2)$ which lead to the same $g(x,c)$, then the Prüfer transformation for these two equations would yield

$$\frac{d\theta}{dx} = \frac{\partial g(x,c)}{\partial x} = \frac{1}{p_{1,i}(x)} \cos^2 g(x,c) - p_{o,i}(x) \sin^2 g(x,c)$$

for i=1,2. Subtracting we get

$$0 \equiv \left(\frac{1}{p_{1,1}(x)} - \frac{1}{p_{1,2}(x)} \right) \cos^2 g(x,c) - \left(p_{o,1}(x) - p_{o,2}(x) \right) \sin^2 g(x,c)$$

for all $x \in \mathcal{J}$. Setting $x = c$ and using the fact that $g(c,c) = 0$ yields $0 \equiv \frac{1}{p_{1,1}(c)} - \frac{1}{p_{1,2}(c)}$ for all $c \in \mathcal{J}$. Therefore $0 \equiv (p_{o,1}(x) - p_{o,2}(x)) \sin^2 g(x,c)$ for $x \in \mathcal{J}$ and $c \in \mathcal{J}$. Since $g(c+\delta,c) \neq 0$ for sufficiently small $|\delta| \neq 0$, it follows that $0 \equiv p_{o,1}(c+\delta) - p_{o,2}(c+\delta)$ for all $c \in \mathcal{J}$ and $|\delta|$ sufficiently small.

Having established uniqueness, one can easily obtain formulas for the coefficients of (7.20). From $g(x,c) = \arctan {}^{u(x,c)}/_{p_1(x)u'(x,c)}$ and the fact that $u(x,c)$ satisfies (7.20) we have

$$\frac{\partial g(x,c)}{\partial x} = \frac{p_1(x)(u'(x;c))^2 - p_o(x)u^2(x;c)}{p_1^2(x)(u'(x;c))^2 + p_o^2(x)u^2(x,c)} .$$

Setting $x = c$ and using the fact that $u(c;c) = 0$ we obtain

$$\left.\frac{\partial g(x,c)}{\partial x}\right|_{x=c} = \frac{1}{p_1(c)} . \tag{7.22}$$

In case $p_o(x) > 0$, an analogous formula can be derived for $p_o(x)$. These formulas can also be obtained from the differential equation which determines $\theta(x)$.

The relationship between (7.22) and Martin's formula follows by solving $w(x;c,\theta) = 0$ for $\theta = h(x,c)$. Since $g(x,c)$ gives the θ-value at x of a solution with a zero at c and $h(x,c)$ gives the θ-value required at c to achieve a zero at x we have

$$\left.\frac{\partial h(x,c)}{\partial x}\right|_{x=c} = - \left.\frac{\partial g(x,c)}{\partial x}\right|_{x=c} = - \frac{1}{p_1(c)} .$$

Also $h(x,c)$ and $f(c,\theta)$ are related by the fact that they are solutions of

$w(x;c,h) = 0$ and $w(f;c,\theta)$, respectively. Therefore for fixed c they are inverse functions, and we have

(7.23)
$$\frac{\partial f}{\partial \theta}\bigg|_{\theta=0} = \left[\frac{\partial h}{\partial x}\bigg|_{x=c} \right]^{-1} = -p_1(c)$$

which is precisely Martin's formula for $p_1(x)$.

An advantage of the development pursued here is that it allows direct generalization to the selfadjoint equation

(7.24)
$$\sum_{k=0}^{n} (-1)^k \left(p_k(x)u^{(k)}\right)^{(k)} = 0.$$

For simplicity we assume $p_k(x)$ is of class C^k and that $p_n(x) > 0$ on $(-\infty,\infty)$. Using the substitutions described earlier, we represent (7.24) in the vector analogue of (7.5),

(7.25)
$$\frac{d\underline{v}}{dx} = G\underline{z} \quad ; \quad \frac{d\underline{z}}{dx} = F\underline{v}$$

where $F(x)$ and $G(x)$ are symmetric continuous matrix functions and $G(x)$ is non-negative definite on $(-\infty,\infty)$.

Corresponding to (7.21) we define

(7.26)
$$B(\underline{v},\underline{z},c,Q) = C(x;c,Q)\underline{v}(x) - S(x;c,Q)\underline{z}(x),$$

where $C(x;c,Q)$ and $S(x;c,Q)$ are defined by (5.4). We shall use the equation $B(\underline{v},\underline{z},c,Q) = 0$ to describe the oscillatory behavior solutions of (7.24) in analogy to the second order case.

In addition to (7.25) we shall be interested in the conjoined matrix solution $V(x;c)$, $Z(x;c)$ of the corresponding matrix system

$$\frac{dV}{dx} = GZ \quad ; \quad \frac{dZ}{dx} = FV$$

(7.27)

$$V(c) = 0 \quad ; \quad Z(c) = I .$$

If $u(x;c,\underline{z}_0)$ denotes the solution of (7.24) which is determined by the initial conditions

(7.28) $$\underline{v}(c) = \underline{0} \quad ; \quad \underline{z}(c) = \underline{z}_0$$

then the corresponding vector solution $\underline{v}(x;c,\underline{z}_0)$ $\underline{z}(x;c,\underline{z}_0)$ of (7.25) has the representation

$$\underline{v}(x;c,\underline{z}_0) = V(x;c)\underline{z}_0 \quad ; \quad \underline{z}(x;c,\underline{z}_0) = Z(x;c)\underline{z}_0 .$$

Therefore $B(\underline{v},\underline{z},c,Q) = 0$ for all solutions of (7.25) with initial conditions of the form (7.28) if and only if

(7.29) $$C(x;c,Q)V(x;c) - S(x;c,Q)Z(x;c) = 0.$$

We seek to show that if (7.29) is satisfied on an interval $\mathcal{J}$, then the coefficients of (7.24) are determined on $\mathcal{J}$.

Let $\tilde{S}(x;c,\tilde{Q})$, $\tilde{C}(x;c,\tilde{Q})$ be matrix solutions of

$$\frac{d\tilde{S}}{dx} = \tilde{Q}(x)\tilde{C} \quad ; \quad \frac{d\tilde{C}}{dx} = -\tilde{Q}(x)\tilde{S}$$

(7.30)

$$\tilde{S}(c) = 0 \quad ; \quad \tilde{C}(c) = I$$

where $\tilde{Q}(x)$ is a continuous symmetric matrix function on $\mathcal{J}$.

<u>7.16 Lemma</u>. <u>If</u> $C\tilde{S}^* - S\tilde{C}^* = 0$ <u>on</u> $\mathcal{J}$, <u>then</u> $S\tilde{S}^* + C\tilde{C}^* = I$ <u>on</u> $\mathcal{J}$.

<u>Proof</u>. Differentiating and making use of (5.4) and (7.30) yields

$$\frac{d}{dx}(S\tilde{S}^*+C\tilde{C}^*) = Q(C\tilde{S}^*-S\tilde{C}^*) + (S\tilde{C}^*-C\tilde{S}^*)\tilde{Q} = 0.$$

Therefore $S\tilde{S}^* + C\tilde{C}^*$ = constant, and from the initial conditions it follows that this constant is the identity.

7.17 Lemma. *If* $C\tilde{S}^* - S\tilde{C}^* = 0$ *on* $\mathcal{I}$, *then* $Q = \tilde{Q}$ *on* $\mathcal{I}$.

Proof. Differentiating and making use of (5.4) and (7.30) yields

$$\frac{d}{dx}(C\tilde{S}^*-S\tilde{C}^*) = (C\tilde{C}^*+S\tilde{S}^*)\tilde{Q} - Q(C\tilde{C}^*+S\tilde{S}^*) = 0$$

on $\mathcal{I}$. By Lemma 7.16, $Q = \tilde{Q}$ on $\mathcal{I}$.

7.18 Theorem. *If* (7.29) *is satisfied for all* $x \in \mathcal{I}$, *then the coefficients of* (7.24) *are determined on* $\mathcal{I}$.

Proof. Using the generalized trigonometric functions $\tilde{S}$, $\tilde{C}$ to represent $Y(x;c)$, $Z(x;c)$ on $\mathcal{I}$, we have

$$Y(x;c) = \tilde{S}^*(x;c,\tilde{Q})\tilde{R}(x;c)$$

$$Z(x;c) = \tilde{C}^*(x;c,\tilde{Q})\tilde{R}(x;c)$$

where $\tilde{Q}(x;c)$ satisfies

$$\tilde{Q}(x,c) = \tilde{C}(x;c,\tilde{Q})G(x)\tilde{C}^*(x;c,\tilde{Q}) - \tilde{S}(x;c,\tilde{Q})F(x)\tilde{S}^*(x;c,\tilde{Q})$$

and $\tilde{R}(x;c)$ is nonsingular on $\mathcal{I}$. Thus (7.29) implies that $C\tilde{S}^* - S\tilde{C}^* = 0$ on $\mathcal{I}$, and it follows from Lemma 7.17 that $Q = \tilde{Q}$ on $\mathcal{I}$. Therefore Q satisfies the functional equation

$$Q(x,c) = C(x;c,Q)G(x)C^*(x;c,Q) - S(x;c,Q)F(x)S^*(x;c,Q) \,. \tag{7.31}$$

If there were two sets of coefficient matrices $F_i(x)$ and $G_i(x)$ $(i=1,2)$ satisfying (7.31) then by subtraction we would have

$$CG_1C^* - SF_1S^* \equiv CG_2C^* - SF_2S^*$$

for all $x, c \in \mathcal{I}$. Setting $x = c$ and using the fact that $S(c;c,Q) = 0$, $C(c;c,Q) = I$, we conclude that $G_1 \equiv G_2$. Therefore

$$S(x;c,Q)F_1(x)S^*(x;c,Q) \equiv S(x;c,Q)F_2(x)S^*(x;c,Q)$$

for all $x, c \in \mathcal{I}$. By a theorem of Etgen [1; Theorem 1.3], it follows that there exists $\delta > 0$ such that $S(x;c,Q)$ is nonsingular for $|x-c| < \delta$. Therefore the last identity implies $F_1(x) \equiv F_2(x)$ and concludes the proof.

The formulas for the coefficients also have analogues in this more general setting. For example, (7.23) can be generalized to

$$Q(c,c) = G(c)$$

by setting $x = c$ in (7.31).

7. Strongly Elliptic Systems

In much the same way as the scalar theory of Chapters 1 and 2 generalizes to elliptic equations, the oscillation theory for Hamiltonian systems generalizes to a class of strongly elliptic systems. Such a generalization based on a Picone-type identity was developed by the author in [13]. A similar development based on a variational approach was given by Swanson [6]. Earlier work in this area was done by Kuks [1] and Bochenek [1].

We consider two strongly elliptic systems of the form

$$\sum_{i,j=1}^{M} \frac{\partial}{\partial x_i}\left(p_{ij}(x)\frac{\partial U}{\partial x_j}\right) + p_o(x)U = 0, \tag{7.32}$$

and

$$-\sum_{i,j=1}^{M} \frac{\partial}{\partial x_i}\left(P_{ij}(x)\frac{\partial V}{\partial x_j}\right) + P_o(x)V = 0, \tag{7.33}$$

where the p_{ij}, P_{ij}, p_o, P_o, U, and V are to be sufficiently regular real $N\times N$ matrices defined in a domain $G \subset E^M$. The ellipticity requirements for (7.32) and (7.33) can be formulated in terms of matrices $\wp$ and Π of order MN which are defined by

$$(\wp)_{ij} = p_{ij} \quad ; \quad (\Pi)_{ij} = P_{ij} .$$

We require that $\wp$ and Π are symmetric and nonnegative definite and that p_o and P_o are symmetric in G.

Our oscillation theory will apply to conjoined solutions of (7.33) satisfying

$$V^* \sum_{j=1}^{M} P_{ij} \frac{\partial V}{\partial x_j} = \left(\sum_{j=1}^{M} P_{ij} \frac{\partial V}{\partial x_j} \right)^* V .$$

If V(x) is a nonsingular conjoined solution of (7.33) and U(x) satisfies (7.32), then the following identity can be established:

$$\sum_i \frac{\partial}{\partial x_i} \left[U^* \sum_j p_{ij} \frac{\partial U}{\partial x_j} - U^* \sum_j P_{ij} \frac{\partial V}{\partial x_j} V^{-1} U \right]$$

$$(7.34) \qquad = U^*(P_o - p_o)U + \sum_{i,j} \frac{\partial U^*}{\partial x_i} (p_{ij} - P_{ij}) \frac{\partial U}{\partial x_i}$$

$$+ \sum_{i,j} \left[\frac{\partial U^*}{\partial x_i} - U^* V^{-1*} \frac{\partial V^*}{\partial x_i} \right] P_{ij} \left[\frac{\partial U}{\partial x_j} - \frac{\partial V}{\partial x_j} V^{-1} U \right] .$$

The matrix U(x) can also be replaced by a vector valued solution of (7.32).

While the identity (7.34) leads to a formal development much like that for scalar elliptic equations, the theory suffers from a major defect: the existence of conjoined solutions of (7.33) is in general difficult to establish. Kuks [1] gives some very special cases of (7.33) in which conjoined solutions are known to exist.

As an example of how such systems arise in oscillation theory, we consider a generalization of the following variant of the Sturm-Picone theorem. If u(x) and v(x) are solutions of $-(p_1(x)u')' + p_o(x)u = 0$ and $-(P_1(x)v')' + P_o(x)v = 0$, respectively, with $P_o(x) \leq p_o(x)$ and $0 < P_1(x) < p_1(x)$ on an interval $\mathcal{J}$, then Theorem 1.3 asserts that the zeros of v(x) separate the zeros of u(x). However, from the Prüfer representation for these equations it follows that the zeros of u'(x) also separate the zeros of v'(x), and the question arises as to whether this latter property also has an analogue for elliptic equations (see Kreith and Travis [2] and Swanson [7]).

We consider the elliptic equations

$$(7.35) \qquad -\sum_{i,j=1}^{n} \frac{\partial}{\partial x_i}\left(a_{ij}(x)\frac{\partial u}{\partial x_j}\right) + c(x)u = 0$$

$$(7.36) \qquad -\sum_{i,j=1}^{n} \frac{\partial}{\partial x_i}\left(A_{ij}(x)\frac{\partial v}{\partial x_j}\right) + C(x)v = 0$$

and suppose (7.35) has a nontrivial solution $u(x)$ whose gradient vanishes on the boundary of a bounded domain $G \subset E^n$. In analogy with the result stated above, one might hope to establish that the conditions

$$(7.37) \qquad C(x) \leq c(x) \quad \text{in } \overline{G}$$

$$(7.38) \qquad 0 < \Sigma A_{ij}(x)\xi_i\xi_j < \Sigma a_{ij}(x)\xi_i\xi_j \quad \text{for all } x \in \overline{G} \text{ and all real n-tuples } \xi_1,\cdots,\xi_n$$

would assure that every solution $v(x)$ of (7.36) has a critical point in $\overline{G}$. However it is easy to construct examples where $\underline{\nabla}v(x)$ fails to have a zero in $\overline{G}$ while (7.37) and (7.38) are satisfied.

To formulate a valid proposition we make the additional assumption $c(x) > 0$ and $C(x) > 0$ in $\overline{G}$ and consider the systems

$$(7.39) \qquad \underline{\nabla}\left(\frac{1}{c(x)}\,\underline{\nabla}\cdot\underline{u}\right) + p_o\underline{u} = 0$$

$$(7.40) \qquad \underline{\nabla}\left(\frac{1}{C(x)}\,\underline{\nabla}\cdot\underline{v}\right) + P_o\underline{v} = 0$$

where $p_o = (a_{ij})^{-1}$, $P_o = (A_{ij})^{-1}$, and $\underline{u}$ and $\underline{v}$ are column vectors whose i-th components are given by

$$-\sum_{j=1}^{n} a_{ij}\frac{\partial u}{\partial x_j} \quad \text{and} \quad -\sum_{j=1}^{n} A_{ij}\frac{\partial v}{\partial x_j}, \text{ respectively.}$$

Then (7.39) and (7.40) are the reciprocal systems of (7.35) and (7.36), respectively, and can be written

(7.41) $$-\sum_{i,j=1} \frac{\partial}{\partial x_i}\left(p_{ij}(x)\frac{\partial \underline{u}}{\partial x_j}\right) + p_o(x)\underline{u} = 0$$

(7.42) $$-\sum_{i,j=1}^{n} \frac{\partial}{\partial x_i}\left(P_{ij}(x)\frac{\partial \underline{v}}{\partial x_j}\right) + P_o(x)\underline{v} = 0$$

where p_{ij} and P_{ij} are $n\times n$ matrices with $\frac{1}{c(x)}$ and $\frac{1}{C(x)}$ in the (i,j)-th block and zeros elsewhere. It is readily verified that the $n^2\times n^2$ matrix whose (i,j)-th block is p_{ij} is symmetric and nonnegative definite, giving rise to the quadratic form

$$\frac{1}{c(x)}(\eta_1 + \eta_{n+2} + \eta_{2n+3} + \cdots + \eta_{n^2})^2 .$$

A similar statement is valid for the $n^2 \times n^2$ matrix composed of the P_{ij}. Thus the following result follows from an application of the comparison theorem for strongly elliptic systems.

7.19 Theorem. Let (7.35) have a nontrivial solution u(x) whose gradient vanishes on the boundary of a bounded domain G. If (7.37) and (7.38) are satisfied in $\overline{G}$, then every prepared matrix solution of

$$\sum_{i,j=1}^{n} \frac{\partial}{\partial x_i}\left(P_{ij}(x)\frac{\partial V}{\partial x_j}\right) + P_o(x)V = 0$$

is singular in $\overline{G}$.

8. A Vector Green's Transform

The Green's transform used to establish nonoscillation criteria for complex second order equations has a generalization to vector differential equations which allows one to establish analogous results for the complex equation

$$(7.41) \qquad \ell u \equiv \sum_{k=0}^{n} (-1)^k (p_k(x) u^{(k)})^{(k)} = 0$$

where $p_k(x)$ is a complex valued function of class C^k and $p_n(x) \neq 0$ on an interval $\mathscr{I}$.

We consider the vector representation

$$(7.42) \qquad \begin{aligned} \underline{u}' &= a(x)\underline{u} + b(x)\underline{w} \\ \underline{w}' &= c(x)\underline{u} + a^*(x)\underline{w} \end{aligned}$$

where $\underline{u}$, $\underline{w}$, and the coefficient matrices are defined as for the real equation (7.2). Using the asterisk to define the complex conjugate transpose (adjoint), it follows from (7.42) that

$$\frac{d}{dx}[\underline{u}^*\underline{w}] = \underline{w}^* b^*(x) \underline{w} + \underline{u}^* c(x) \underline{u}.$$

If $\alpha < \beta$ are points of $\mathscr{I}$, then

$$\Big[\underline{u}^*\underline{w}\Big]_\alpha^\beta = \int_\alpha^\beta \Big[\underline{w}^* b^*(x) \underline{w} + \underline{u}^* c(x) \underline{u}\Big] dx$$

$$= \int_\alpha^\beta \Big[\frac{1}{\overline{p_n(x)}} |u_n|^2 + \sum_{k=0}^{n-1} p_k(x) |u^{(k)}|^2 \Big] dx .$$

Considering the special case $p_n(x) > 0$ and taking real and imaginary parts yields

$$\mathrm{Re} \Big[\underline{u}^*\underline{w}\Big]_\alpha^\beta = \int_\alpha^\beta \Big[\frac{1}{p_n(x)} |u_n|^2 + \sum_{k=0}^{n-1} \mathrm{Re}\; p_k(x) |u^{(k)}|^2 \Big] dx$$

$$\mathrm{Im} \Big[\underline{u}^*\underline{w}\Big]_\alpha^\beta = \int_\alpha^\beta \Big[\sum_{k=0}^{n-1} \mathrm{Im}\; p_k(x) |u^{(k)}|^2 \Big] dx.$$

These equations readily yield the following.

7.20 Theorem. *If* $p_n(x) > 0$ *and*

(i) $\operatorname{Re} p_k(x) \geq 0$

or

(ii) $\operatorname{Im} p_k(x) \geq 0$ (*but not all identically zero*)

or

(iii) $\operatorname{Im} p_k(x) \leq 0$ (*but not all identically zero*)

for $0 \leq k \leq n-1$ *and all* $x \in (\alpha,\beta)$, *then* (7.41) *is disconjugate on* $[\alpha,\beta]$.

CHAPTER 8

FOURTH ORDER DIFFERENTIAL EQUATIONS

1. Sturmian Theorems

Many oscillation properties have been established for fourth order differential equations which have no direct or obvious generalizations to the case of equations of even order considered in Chapter 7. Without attempting to cover the vast literature on this subject, we shall present several results of this sort which are closely related to the topics previously presented.

Diaz and Dunninger [1] have considered functions u(x) and v(x) which are nontrivial solutions of

$$\Delta^2 u + p(x)u = 0 \tag{8.1}$$

$$\Delta^2 v + P(x)v = 0 \tag{8.2}$$

in a domain $D \subset E^n$. Multiplying (8.1) and (8.2) by v and u, respectively, and subtracting yields

$$u\Delta^2 v - v\Delta^2 u = (p-P)uv.$$

An application of Green's theorem yields

$$\iint_D (p-P)uv\, dx = \int_{\partial D}\left[u\,\frac{\partial(\Delta v)}{\partial \nu} - v\,\frac{\partial(\Delta u)}{\partial \nu} + \Delta u\,\frac{\partial v}{\partial \nu} - \Delta v\,\frac{\partial u}{\partial \nu}\right] d\bar{x}\ , \tag{8.3}$$

where $\frac{\partial}{\partial n}$ denotes the exterior normal derivative. A direct consequence of (8.3) is the following Sturmian comparison theorem.

8.1 Theorem. *Suppose there exists a nontrivial solution u(x) of (8.1) which satisfies $u(x) \geq 0$ in D and any one of the following boundary conditions:*

(i) $u = \Delta u = 0$ *on* ∂D

(ii) $u = \frac{\partial u}{\partial \nu} = 0$ <u>on</u> ∂D

(iii) $\frac{\partial u}{\partial \nu} = \frac{\partial(\Delta u)}{\partial \nu} = 0$ <u>on</u> ∂D

(iv) $\Delta u = \frac{\partial(\Delta u)}{\partial \nu} = 0$ <u>on</u> ∂D.

If $p(x) \geq P(x)$ *(but not identically equal) in* D *then every solution* $v(x)$ *of* (8.2) *satisfying the same boundary condition as* $u(x)$ *has a zero in* D.

This basic result allows numerous variations in which some of the boundary conditions on $v(x)$ may be omitted.

Another special result due to the author [14] applies to solutions $u(x)$ and $v(x)$ if

$$u^{(iv)} + p(x)u = 0 \tag{8.4}$$

$$v^{(iv)} + P(x)v = 0 \tag{8.5}$$

on an interval $\mathcal{I} = (x_1, x_2)$ which satisfy boundary conditions of the form

$$\sigma_{ij}u^{(j-1)}(x_i) + (-1)^{i+j}u^{(4-j)}(x_i) = 0 \quad ; \quad i,j=1,2 \quad , \tag{8.6}$$

$$\tau_{ij}v^{(j-1)}(x_i) + (-1)^{i+j}v^{(4-j)}(x_i) = 0 \quad ; \quad i,j=1,2 \quad , \tag{8.7}$$

respectively. If the σ_{ij} and τ_{ij} are all positive, then (8.6) and (8.7) are called "boundary conditions of constraint". It can be shown that the Green's function for the equation $u^{(iv)} = f(x)$ is positive when boundary conditions of constraint are imposed, and in this case the theory of positive operators developed in Chapter 5 applies. In this instance this theory yields the following result.

8.2 Theorem. *Let* $u(x)$ *and* $v(x)$ *be nontrivial solutions of* (8.4) *and* (8.5) *satisfying* (8.6) *and* (8.7), *respectively. If*

(i) $0 > p(x) \geq P(x)$ <u>in</u> $\mathcal{I}$

(ii) $\sigma_{ij} > \tau_{ij} > 0$ *for* $i,j=1,2$

then $v(x)$ *has a zero in* (x_1,x_2).

2. Generalized Picone Identities

The Picone identity (1.4) also allows some generalizations to fourth order equations which do not require the Hamiltonian systems employed in Chapter 7. In developing such identities for the fourth order equation directly one circumvents the requirement for conjoined solutions, and this fact may be of advantage in dealing with nonselfadjoint problems.

If oscillation is to be measured in terms of conjugate points, a very natural generalization of (1.4) can be based on a treatment of the extended Legendre condition for quadratic functionals of the form

$$\int_\alpha^\beta \left[P_2(x)y''^2 + P_1(x)y'^2 + P_o(x)y^2\right] dx$$

due to Leighton [4]. Letting $v_1(x)$ and $v_2(x)$ be linearly independent solutions of

$$(P_2(x)v'')'' - (P_1(x)v')' + P_o(x)v = 0, \tag{8.8}$$

where $P_k(x)$ is of class C^k and $P_2(x)$ is positive on $\mathcal{I}$, we define

$$\sigma = v_1v_2' - v_1'v_2$$

$$\tau = v_1'v_2'' - v_1''v_2' .$$

It can then be shown that following is an identity in u, u', u'' whenever $\sigma \neq 0$:

$$\frac{d}{dx}\left[P_2 \frac{\sigma'}{\sigma}u'^2 - 2P_2 \frac{\tau}{\sigma}uu' + \frac{(P_2\tau)'}{\sigma}u^2\right]$$
$$= P_2u''^2 + P_1u'^2 + P_ou^2 - P_2(u'' - \frac{\sigma'}{\sigma}u' + \frac{\tau}{\sigma}u)^2 . \tag{8.9}$$

If in addition u(x) is a solution of

$$\ell u \equiv (p_2(x)u'')'' - (p_1(x)u')' + p_o(x)u = 0 \tag{8.10}$$

then

$$\frac{d}{dx}[-(p_2u'')'u' + p_2u''u' + p_1u'u] = p_2u''^2 + p_1u'^2 + p_ou^2. \tag{8.11}$$

Subtracting (8.9) from (8.11) yields

$$\frac{d}{dx}\Big[-(p_2u'')'u' + p_2u''u' + p_1u'u - P_2\frac{\sigma'}{\sigma}u'^2 + 2P_2\frac{\tau}{\sigma}uu' - \frac{(P_2\tau)'}{\sigma}u^2\Big] = (p_2-P_2)u''^2 + (p_1-P_1)u'^2 + (p_o-P_o)u^2 + P_2(u'' - \frac{\sigma'}{\sigma}u' + \frac{\tau}{\sigma}u)^2 . \tag{8.12}$$

Since the zeros of $\sigma(x)$ correspond to the conjugate points of α (with respect to ℓ), the identity (8.12) yields directly a proof of Theorem 7.4 for n=2. In this sense the identity (8.12) is the natural generalization of the classical Picone identity (1.4) to fourth order equations.

Two questions naturally arise:

(i) Does (8.12) allow a generalization to nonselfadjoint equations analogous to the generalizations of (1.4) which are contained in Chapter 2?

(ii) Does (8.12) allow a generalization to selfadjoint equations of arbitrary even order?

The answer to these questions appears to be unknown.

Another Picone-type identity has been established by Dunninger [2] for fourth order elliptic equations of the form

$$\ell u \equiv \Delta(p_1(x)\Delta u) + p_o(x)u = 0 \tag{8.13}$$

$$Lv \equiv \Delta(P_1(x)\Delta v) + P_o(x)v = 0 \tag{8.14}$$

in a domain $D \subset E^n$.

Using Green's theorem one establishes the identities

$$\int_{\partial D}\left[u\frac{\partial(p_1\Delta u)}{\partial\nu} - p_1\Delta u\frac{\partial u}{\partial\nu}\right]d\bar{x} = \iint_D u\,\ell u\,dx - \iint_D\left[p_1(\Delta u)^2 + p_o u^2\right]dx$$

$$\int_{\partial D}\left[P_1\Delta v\frac{\partial}{\partial\nu}\left(\frac{u^2}{v}\right) - \frac{u^2}{v}\frac{\partial(P_1\Delta v)}{\partial\nu}\right]d\bar{x} = \iint_D\left[P_1\Delta v\Delta\left(\frac{u^2}{v}\right) + P_o u^2\right]dx - \iint_D\frac{u^2}{v}Lv\,dx$$

whenever $v(x) = 0$. Adding these and making use of the expressions

$$P_1\Delta v\Delta\left(\frac{u^2}{v}\right) = P_1(\Delta u)^2 + 2P_1\frac{\Delta v}{v}\left|\nabla u - \frac{u}{v}\nabla v\right|^2 - P_1\left(\Delta u - \frac{u}{v}\Delta v\right)^2$$

$$\frac{\partial}{\partial\nu}\left(\frac{u^2}{v}\right) = 2\frac{u}{v}\frac{\partial u}{\partial\nu} - \frac{u^2}{v^2}\frac{\partial v}{\partial\nu}$$

yields

$$\begin{aligned}
&\int_{\partial D}\frac{u}{v}\left[v\frac{\partial(p_1\Delta u)}{\partial\nu} - u\frac{\partial(P_1\Delta v)}{\partial\nu}\right]d\bar{x} + \int_{\partial D}P_1\frac{\Delta v}{v}\left[\frac{u}{v}\left(v\frac{\partial u}{\partial\nu} - u\frac{\partial v}{\partial\nu}\right)\right]d\bar{x}\\
&+\int_{\partial D}\frac{1}{v}\frac{\partial u}{\partial\nu}(P_1 u\Delta v - p_1 v\Delta u)\,d\bar{x} = \iint_D\left[(P_1 - p_1)(\Delta u)^2 + (P_o - p_o)u^2\right]dx\\
&+\iint_D\left[2P_1\frac{\Delta v}{v}\left|\nabla u - \frac{u}{v}\nabla v\right|^2 - P_1\left(\Delta u - \frac{u}{v}\Delta v\right)^2\right]dx\\
&+\iint_D\frac{u}{v}(v\,\ell u - uLv)\,dx
\end{aligned} \tag{8.15}$$

As an example of the type of theorem which follows from this identity we quote the following.

8.3 Theorem. *Suppose* u(x) *is a nontrivial solution of* (8.13) *in* D *satisfying* $u = \frac{\partial u}{\partial \nu} = 0$ *on* ∂D. *If*

(i) $0 < P_1(x) \leq p_1(x)$ *in* D

(ii) $P_o(x) \leq p_o(x)$ *in* D

then every solution of (8.14) *which satisfies*

(iii) $\Delta v < 0$ *in* D

(iv) $v(x) > 0$ *for some* $x \in D$

has a zero in D *or else is a constant multiple of* u(x).

REFERENCES

AHLBRANDT, C. D.:

[1] Disconjugacy criteria for selfadjoint differential systems, J. Differential Equations 6(1969), 271-295.

ALLEGRETTO, W.:

[1] A comparison theorem for nonlinear operators, Annali Scuola Norm. Sup. Pisa 25(1971), 41-46.

[2] Eigenvalue comparison and oscillation criteria for elliptic operators, J. London Math. Soc. 3(1971), 571-575.

BARRETT, J.:

[1] Oscillation theory of ordinary differential equations, Advances in Math. 3(1969), 415-509.

[2] A Prüfer transformation for matrix differential equations, Proc. Amer. Math. Soc. 8(1957), 510-518.

[3] Second order complex differential equations with a real independent variable, Pacific J. Math. 8(1958), 187-200.

BARTA, J.:

[1] Sur la vibration fondamentale dúne membrane, C. R. Acad. Sci. Paris 204(1937), 472-473.

BENSON, D. and K. KREITH:

[1] On abstract Prüfer transformations, Proc. Amer. Math. Soc. 26(1970), 137-140.

BIRKHOFF, G. and G. C. ROTA:

[1] Ordinary Differential Equations (Ginn and Company, Boston, 1962).

BOCHENEK, J.:

[1] On eigenvalues and eigenfunctions of strongly elliptic systems of differential equations of second order, Prace Mat. 12(1968), 171-182.

CLARK, C. and C. A. SWANSON:

[1] Comparison theorems for elliptic differential equations, Proc. Amer. Math. Soc. 16(1965), 886-890.

DIAZ, J. B. and D. R. DUNNINGER:

[1] Sturm separation and comparison theorems for a class of fourth order ordinary and partial differential equations, to appear, Applicable Analysis.

DIEUDONNÉ, J.:

[1] Foundations of Modern Analysis (Academic Press, New York, 1960).

DUNFORD, N. and J. SCHWARTZ:

[1] Linear Operators II (Wiley, New York, 1963).

DUNNINGER, D. R.:

[1] A Picone identity for non-self-adjoint elliptic operators, Atti Accad. Naz. Lincei Rend. 48(1970), 133-139.

[2] A Picone integral identity for a class of fourth order elliptic differential inequalities, Atti Accad. Naz. Lincei Rend. 50(1971), 630-641.

ETGEN, G. J.:

[1] Oscillatory properties of certain nonlinear matrix differential systems, Trans. Amer. Math. Soc. 122(1966), 289-310.

[2] A note on trigonometric matrices, Proc. Amer. Math. Soc. 17(1966), 1226-1232.

FINK, A. M.:

[1] A converse problem in differential equations, SIAM J. Appl. Math. 21(1971), 355-361.

HEADLEY, V. B. and C. A. SWANSON:

[1] Oscillation criteria for elliptic equations, Pacific J. Math. 29(1968), 501-506.

HILLE, E.:

[1] Lectures on Ordinary Differential Equations (Addison-Wesley, Reading, Mass., 1969).

INCE, E. L.:

[1] Ordinary Differential Equations (Dover, New York, 1956).

JENTZCH, R.:

[1] Über Integralgleichungen mit positivem Kern, J. Math. von Crelle, 141(1912), 235-244.

KAMKE, E.:

[1] A new proof of Sturm's comparison theorems, Amer. Math. Monthly 46(1939), 417-421.

[2] Über Sturm's Vergleichssätze für homogene lineare differentialgleichungen zweiter Ordnung und Systeme von zwei Differentialgleichungen erster ordnung, Math. Z. 47(1942), 788-795.

KNESER, A.:

[1] Untersuchungen über die reelen Nullstellen der Integrale linearer Differentialgleichungen, Math. Ann. 42(1893), 409-435.

KRASNOSELSKII, M.:

[1] Positive Solutions of Operator Equations (Noordhoff, Groningen, 1964).

KREITH, K.:

[1] A strong comparison theorem for selfadjoint elliptic equations, Proc. Amer. Math. Soc. 19(1968), 989-990.

[2] A class of comparison theorems for nonselfadjoint elliptic equations, Proc. Amer. Math. Soc. 29(1971), 547-552.

[3] Oscillation theorems for elliptic equations, Proc. Amer. Math. Soc. 15(1964), 341-344.

[4] Sturmian theorems for hyperbolic equations, Proc. Amer. Math. Soc. 22(1969), 277-281.

[5] Sturmian theorems for characteristic initial value problems, Accad. Naz. dei Lincei, 47(1969), 139-144.

[6] Complementary bounds for eigenvalues, Atti Accad. Lincei 66(1969), 164-165.

KREITH, K. (Continued):

[7] Criteria for positive Green's functions, Illinois J. Math. 12(1968), 475-478.

[8] Sturmian theorems and positive resolvents, Trans. Amer. Math. Soc. 139(1969), 319-327.

[9] An abstract oscillation theorem, Proc. Amer. Math. Soc. 19(1968), 17-20.

[10] Disconjugacy criteria for nonselfadjoint differential equations of even order, Canadian J. Math. 23(1971), 644-652.

[11] A Prüfer transformation for nonselfadjoint systems, Proc. Amer. Math. Soc. 31(1972), 147-151.

[12] Oscillation criteria for nonlinear matrix differential equations, Proc. Amer. Math. Soc. 26(1970), 270-272.

[13] A Picone identity for strongly elliptic systems, Duke Math. J. 38(1971), 473-481.

[14] Comparison theorems for constrained rods, SIAM Review 6(1964), 31-36.

KREITH, K. and C. TRAVIS:

[1] Oscillation criteria for selfadjoint elliptic equations, Pacific J. Math., 41(1972), 743-753.

[2] On a comparison theorem for strongly elliptic systems, J. Differential Equations 10(1971), 173-178.

KUKS, L. M.:

[1] Sturm's theorem and oscillation of solutions of strongly elliptic systems, Dokl. Akad. Nauk. SSSR, 42(1962), 32-35.

LADAS, G.:

[1] Connection between oscillation and spectrum for selfadjoint differential operators of order 2n, Comm. on Pure and Appl. Math. 22(1969), 561-585.

LEIGHTON, W.:

[1] On selfadjoint differential equations of second order, J. London Math. Soc. 27(1952), 37-47.

[2] A substitute for the Picone formula, Bull. Amer. Math. Soc. 55(1949), 325-328.

[3] Comparison theorems for linear differential equations of second order, Proc. Amer. Math. Soc. 13(1962), 603-610.

[4] Quadratic functionals of second order, Trans. Amer. Math. Soc. 151(1970), 309-322.

LEVIN, A. Ju.:

[1] A comparison principle for second-order differential equations, Soviet Math. Dokl. 1(1960), 1313-1316.

MARTIN, A. D.:

[1] An inverse Sturm-Liouville problem, Duke Math. J. 26(1959), 455-466.

NEHARI, Z.:

[1] Oscillation criteria for second-order linear differential equations, Trans. Amer. Math. Soc. 85(1957), 428-445.

NOUSSAIR, E. S.:

[1] Oscillation theory of elliptic equations of order 2m, J. Differential Equations 10(1971), 100-111.

PICARD, E.:

[1] Lecons sur quelques problèmes aux limites de la Théorie des équations différentielles, (Paris, 1930).

PICONE, M.:

[1] Sui valori eccezionali di un parametro da cui dipende un'equazione differenziale lineare ordinaria del secondo ordine, Ann. Scuola Norm. Pisa 11(1910), 1-141.

PICONE, M. (Continued):

[2] Un teorema sulle soluzioni delle equazioni lineari ellitiche autoaggiunte alle derivate parziali del secondo-ordine, Atti Accad. Lincei 20(1911), 213-219.

PROTTER, M. H.:

[1] A comparison theorem for elliptic equations, Proc. Amer. Math. Soc. 10(1959), 296-299.

PROTTER, M. H. and H. F. WEINBERGER:

[1] On the spectrum of general second order operators, Bull. Amer. Math. Soc., 72(1966), 251-255.

PRÜFER, H.:

[1] Neue Herleitung der Sturm-Liouvilleschen Reihenentwicklung stetiger Funktionen, Math. Ann. 95(1926), 499-518.

REID, W. T.:

[1] Ordinary Differential Equations (Wiley, New York, 1971).

[2] Riccati Differential Equations (Academic Press, New York, 1972).

[3] A Prüfer transformation for differential systems, Pacific J. Math. 8(1958), 575-584.

[4] Generalized polar coordinate transformations for differential systems, Rocky Mountain J. Math. 1(1971), 383-406.

[5] A class of monotone Riccati matrix differential operators, Duke Math. J. 32(1965), 689-696.

[6] Variational methods and boundary value problems for ordinary linear differential systems, "The Proceedings of the Japan-United States Seminar on Functional and Differential Equations", 267-299. Benjamin, New York, 1967.

RUTMAN, M.:

[1] Sur les opérateurs totalement continus linéaires laissant invariant un certain cone, Math. Sbornik 8(1940), 77-96.

STURM, C.:

[1] Sur les equations differentielles lineaires du second ordre, J. Math. Pures Appl. 1(1836), 106-186.

SWANSON, C. A.:

[1] Comparison and Oscillation Theory of Linear Differential Equations (Academic Press, New York and London, 1968).

[2] Strong oscillation of elliptic equations in general domain, Canadian Math. Bull., to appear.

[3] A comparison theorem for elliptic differential equations, Proc. Amer. Math. Soc. 17(1966), 611-616.

[4] An identity for elliptic equations with applications, Trans. Amer. Math. Soc. 134(1968), 325-333.

[5] Comparison theorems for elliptic equations on unbounded domains, Trans. Amer. Math. Soc. 126(1967), 278-285.

[6] Comparison theorems for elliptic systems, Pacific J. Math. 33(1970), 445-450.

[7] Remarks on a comparison theorem of Kreith and Travis, J. Differential Equations 11(1972), 624-627.

TAAM, C. T.:

[1] Oscillation theorems, Amer. J. Math. 74(1952), 317-324.

TOMASTIK, E. C.:

[1] Oscillation of nonlinear matrix differential equations of second order, Proc. Amer. Math. Soc. 19(1968), 1427-1431.

TRAVIS, C.:

[1] Comparison of eigenvalues for linear differential equations of order 2n, Trans. Amer. Math. Soc., to appear.

WILLETT, D.:

[1] Classification of second order linear differential equations with respect to oscillation, Advances in Math. 3(1969), 594-623.

WONG, P. K.:

[1] A Sturmian theorem for first order partial differential equations, Trans. Amer. Math. Soc. 166(1972), 225-240.

COPPEL, W. A.:

[1] Disconjugacy, Lecture Notes in Mathematics #220, Springer-Verlag (Berlin), 1971.

GELFAND, I. M. and S. V. FOMIN:

[1] Calculus of Variations, Prentice-Hall (Englewood Cliffs, New Jersey), 1963.

Lecture Notes in Mathematics

Comprehensive leaflet on request

Vol. 146: A. B. Altman and S. Kleiman, Introduction to Grothendieck Duality Theory. II, 192 pages. 1970. DM 18,–

Vol. 147: D. E. Dobbs, Cech Cohomological Dimensions for Commutative Rings. VI, 176 pages. 1970. DM 16,–

Vol. 148: R. Azencott, Espaces de Poisson des Groupes Localement Compacts. IX, 141 pages. 1970. DM 16,–

Vol. 149: R. G. Swan and E. G. Evans, K-Theory of Finite Groups and Orders. IV, 237 pages. 1970. DM 20,–

Vol. 150: Heyer, Dualität lokalkompakter Gruppen. XIII, 372 Seiten. 1970. DM 20,–

Vol. 151: M. Demazure et A. Grothendieck, Schémas en Groupes I. (SGA 3). XV, 562 pages. 1970. DM 24,–

Vol. 152: M. Demazure et A. Grothendieck, Schémas en Groupes II. (SGA 3). IX, 654 pages. 1970. DM 24,–

Vol. 153: M. Demazure et A. Grothendieck, Schémas en Groupes III. (SGA 3). VIII, 529 pages. 1970. DM 24,–

Vol. 154: A. Lascoux et M. Berger, Variétés Kähleriennes Compactes. VII, 83 pages. 1970. DM 16,–

Vol. 155: Several Complex Variables I, Maryland 1970. Edited by J. Horváth. IV, 214 pages. 1970. DM 18,–

Vol. 156: R. Hartshorne, Ample Subvarieties of Algebraic Varieties. XIV, 256 pages. 1970. DM 20,–

Vol. 157: T. tom Dieck, K. H. Kamps und D. Puppe, Homotopietheorie. VI, 265 Seiten. 1970. DM 20,–

Vol. 158: T. G. Ostrom, Finite Translation Planes. IV. 112 pages. 1970. DM 16,–

Vol. 159: R. Ansorge und R. Hass. Konvergenz von Differenzenverfahren für lineare und nichtlineare Anfangswertaufgaben. VIII, 145 Seiten. 1970. DM 16,–

Vol. 160: L. Sucheston, Constributions to Ergodic Theory and Probability. VII, 277 pages. 1970. DM 20,–

Vol. 161: J. Stasheff, H-Spaces from a Homotopy Point of View. VI, 95 pages. 1970. DM 16,–

Vol. 162: Harish-Chandra and van Dijk, Harmonic Analysis on Reductive p-adic Groups. IV, 125 pages. 1970. DM 16,–

Vol. 163: P. Deligne, Equations Différentielles à Points Singuliers Reguliers. III, 133 pages. 1970. DM 16,–

Vol. 164: J. P. Ferrier, Seminaire sur les Algebres Complètes. II, 69 pages. 1970. DM 16,–

Vol. 165: J. M. Cohen, Stable Homotopy. V, 194 pages. 1970. DM 16.–

Vol. 166: A. J. Silberger, PGL_2 over the p-adics: its Representations, Spherical Functions, and Fourier Analysis. VII, 202 pages. 1970. DM 18,–

Vol. 167: Lavrentiev, Romanov and Vasiliev, Multidimensional Inverse Problems for Differential Equations. V, 59 pages. 1970. DM 16,–

Vol. 168: F. P. Peterson, The Steenrod Algebra and its Applications: A conference to Celebrate N. E. Steenrod's Sixtieth Birthday. VII, 317 pages. 1970. DM 22,–

Vol. 169: M. Raynaud, Anneaux Locaux Henséliens. V, 129 pages. 1970. DM 16,–

Vol. 170: Lectures in Modern Analysis and Applications III. Edited by C. T. Taam. VI, 213 pages. 1970. DM 18,–

Vol. 171: Set-Valued Mappings, Selections and Topological Properties of 2^X. Edited by W. M. Fleischman. X, 110 pages. 1970. DM 16,–

Vol. 172: Y.-T. Siu and G. Trautmann, Gap-Sheaves and Extension of Coherent Analytic Subsheaves. V, 172 pages. 1971. DM 16,–

Vol. 173: J. N. Mordeson and B. Vinograde, Structure of Arbitrary Purely Inseparable Extension Fields. IV, 138 pages. 1970. DM 16,–

Vol. 174: B. Iversen, Linear Determinants with Applications to the Picard Scheme of a Family of Algebraic Curves. VI, 69 pages. 1970. DM 16,–

Vol. 175: M. Brelot, On Topologies and Boundaries in Potential Theory. VI, 176 pages. 1971. DM 18,–

Vol. 176: H. Popp, Fundamentalgruppen algebraischer Mannigfaltigkeiten. IV, 154 Seiten. 1970. DM 16,–

Vol. 177: J. Lambek, Torsion Theories, Additive Semantics and Rings of Quotients. VI, 94 pages. 1971. DM 16,–

Vol. 178: Th. Bröcker und T. tom Dieck, Kobordismentheorie. XVI, 191 Seiten. 1970. DM 18,–

Vol. 179: Seminaire Bourbaki – vol. 1968/69. Exposés 347-363. IV. 295 pages. 1971. DM 22,–

Vol. 180: Séminaire Bourbaki – vol. 1969/70. Exposés 364-381. IV, 310 pages. 1971. DM 22,–

Vol. 181: F. DeMeyer and E. Ingraham, Separable Algebras over Commutative Rings. V, 157 pages. 1971. DM 16.–

Vol. 182: L. D. Baumert. Cyclic Difference Sets. VI, 166 pages. 1971. DM 16,–

Vol. 183: Analytic Theory of Differential Equations. Edited by P. F. Hsieh and A. W. J. Stoddart. VI, 225 pages. 1971. DM 20,–

Vol. 184: Symposium on Several Complex Variables, Park City, Utah, 1970. Edited by R. M. Brooks. V, 234 pages. 1971. DM 20,–

Vol. 185: Several Complex Variables II, Maryland 1970. Edited by J. Horváth. III, 287 pages. 1971. DM 24,–

Vol. 186: Recent Trends in Graph Theory. Edited by M. Capobianco/ J. B. Frechen/M. Krolik. VI, 219 pages. 1971. DM 18.–

Vol. 187: H. S. Shapiro, Topics in Approximation Theory. VIII, 275 pages. 1971. DM 22,–

Vol. 188: Symposium on Semantics of Algorithmic Languages. Edited by E. Engeler. VI, 372 pages. 1971. DM 26,–

Vol. 189: A. Weil, Dirichlet Series and Automorphic Forms. V, 164 pages. 1971. DM 16,–

Vol. 190: Martingales. A Report on a Meeting at Oberwolfach, May 17-23, 1970. Edited by H. Dinges. V, 75 pages. 1971. DM 16,–

Vol. 191: Séminaire de Probabilités V. Edited by P. A. Meyer. IV, 372 pages. 1971. DM 26,–

Vol. 192: Proceedings of Liverpool Singularities – Symposium I. Edited by C. T. C. Wall. V, 319 pages. 1971. DM 24,–

Vol. 193: Symposium on the Theory of Numerical Analysis. Edited by J. Ll. Morris. VI, 152 pages. 1971. DM 16,–

Vol. 194: M. Berger, P. Gauduchon et E. Mazet. Le Spectre d'une Variété Riemannienne. VII, 251 pages. 1971. DM 22,–

Vol. 195: Reports of the Midwest Category Seminar V. Edited by J.W. Gray and S. Mac Lane.III, 255 pages. 1971. DM 22,–

Vol. 196: H-spaces – Neuchâtel (Suisse)- Août 1970. Edited by F. Sigrist, V, 156 pages. 1971. DM 16,–

Vol. 197: Manifolds – Amsterdam 1970. Edited by N. H. Kuiper. V, 231 pages. 1971. DM 20,–

Vol. 198: M. Hervé, Analytic and Plurisubharmonic Functions in Finite and Infinite Dimensional Spaces. VI, 90 pages. 1971. DM 16.–

Vol. 199: Ch. J. Mozzochi, On the Pointwise Convergence of Fourier Series. VII, 87 pages. 1971. DM 16,–

Vol. 200: U. Neri, Singular Integrals. VII, 272 pages. 1971. DM 22,–

Vol. 201: J. H. van Lint, Coding Theory. VII, 136 pages. 1971. DM 16,–

Vol. 202: J. Benedetto, Harmonic Analysis on Totally Disconnected Sets. VIII, 261 pages. 1971. DM 22,–

Vol. 203: D. Knutson, Algebraic Spaces. VI, 261 pages. 1971. DM 22,–

Vol. 204: A. Zygmund, Intégrales Singulières. IV, 53 pages. 1971. DM 16,–

Vol. 205: Séminaire Pierre Lelong (Analyse) Année 1970. VI, 243 pages. 1971. DM 20,–

Vol. 206: Symposium on Differential Equations and Dynamical Systems. Edited by D. Chillingworth. XI, 173 pages. 1971. DM 16,–

Vol. 207: L. Bernstein, The Jacobi-Perron Algorithm – Its Theory and Application. IV, 161 pages. 1971. DM 16,–

Vol. 208: A. Grothendieck and J. P. Murre, The Tame Fundamental Group of a Formal Neighbourhood of a Divisor with Normal Crossings on a Scheme. VIII, 133 pages. 1971. DM 16,–

Vol. 209: Proceedings of Liverpool Singularities Symposium II. Edited by C. T. C. Wall. V, 280 pages. 1971. DM 22,–

Vol. 210: M. Eichler, Projective Varieties and Modular Forms. III, 118 pages. 1971. DM 16,–

Vol. 211: Théorie des Matroïdes. Edité par C. P. Bruter. III, 108 pages. 1971. DM 16,–

Please turn over

Vol. 212: B. Scarpellini, Proof Theory and Intuitionistic Systems. VII, 291 pages. 1971. DM 24,–

Vol. 213: H. Hogbe-Nlend, Théorie des Bornologies et Applications. V, 168 pages. 1971. DM 18,–

Vol. 214: M. Smorodinsky, Ergodic Theory, Entropy. V, 64 pages. 1971. DM 16,–

Vol. 215: P. Antonelli, D. Burghelea and P. J. Kahn, The Concordance-Homotopy Groups of Geometric Automorphism Groups. X, 140 pages. 1971. DM 16,–

Vol. 216: H. Maaß, Siegel's Modular Forms and Dirichlet Series. VII, 328 pages. 1971. DM 20,–

Vol. 217: T. J. Jech, Lectures in Set Theory with Particular Emphasis on the Method of Forcing. V, 137 pages. 1971. DM 16,–

Vol. 218: C. P. Schnorr, Zufälligkeit und Wahrscheinlichkeit. IV, 212 Seiten 1971. DM 20,–

Vol. 219: N. L. Alling and N. Greenleaf, Foundations of the Theory of Klein Surfaces. IX, 117 pages. 1971. DM 16,–

Vol. 220: W. A. Coppel, Disconjugacy. V, 148 pages. 1971. DM 16,–

Vol. 221: P. Gabriel und F. Ulmer, Lokal präsentierbare Kategorien. V, 200 Seiten. 1971. DM 18,–

Vol. 222: C. Meghea, Compactification des Espaces Harmoniques. III, 108 pages. 1971. DM 16,–

Vol. 223: U. Felgner, Models of ZF-Set Theory. VI, 173 pages. 1971. DM 16,–

Vol. 224: Revètements Etales et Groupe Fondamental. (SGA 1). Dirigé par A. Grothendieck XXII, 447 pages. 1971. DM 30,–

Vol. 225: Théorie des Intersections et Théorème de Riemann-Roch. (SGA 6). Dirigé par P. Berthelot, A. Grothendieck et L. Illusie. XII, 700 pages. 1971. DM 40,–

Vol. 226: Seminar on Potential Theory, II. Edited by H. Bauer. IV, 170 pages. 1971. DM 18,–

Vol. 227: H. L. Montgomery, Topics in Multiplicative Number Theory. IX, 178 pages. 1971. DM 18,–

Vol. 228: Conference on Applications of Numerical Analysis. Edited by J. Ll. Morris. X, 358 pages. 1971. DM 26,–

Vol. 229: J. Väisälä, Lectures on n-Dimensional Quasiconformal Mappings. XIV, 144 pages. 1971. DM 16,–

Vol. 230: L. Waelbroeck, Topological Vector Spaces and Algebras. VII, 158 pages. 1971. DM 16,–

Vol. 231: H. Reiter, L^1-Algebras and Segal Algebras. XI, 113 pages. 1971. DM 16,–

Vol. 232: T. H. Ganelius, Tauberian Remainder Theorems. VI, 75 pages. 1971. DM 16,–

Vol. 233: C. P. Tsokos and W. J. Padgett. Random Integral Equations with Applications to Stochastic Systems. VII, 174 pages. 1971. DM 18,–

Vol. 234: A. Andreotti and W. Stoll. Analytic and Algebraic Dependence of Meromorphic Functions. III, 390 pages. 1971. DM 26,–

Vol. 235: Global Differentiable Dynamics. Edited by O. Hájek, A. J. Lohwater, and R. McCann. X, 140 pages. 1971. DM 16,–

Vol. 236: M. Barr, P. A. Grillet, and D. H. van Osdol. Exact Categories and Categories of Sheaves. VII, 239 pages. 1971, DM 20,–

Vol. 237: B. Stenström. Rings and Modules of Quotients. VII, 136 pages. 1971. DM 16,–

Vol. 238: Der kanonische Modul eines Cohen-Macaulay-Rings. Herausgegeben von Jürgen Herzog und Ernst Kunz. VI, 103 Seiten. 1971. DM 16,–

Vol. 239: L. Illusie, Complexe Cotangent et Déformations I. XV, 355 pages. 1971. DM 26,–

Vol. 240: A. Kerber, Representations of Permutation Groups I. VII, 192 pages. 1971. DM 18,–

Vol. 241: S. Kaneyuki, Homogeneous Bounded Domains and Siegel Domains. V, 89 pages. 1971. DM 16,–

Vol. 242: R. R. Coifman et G. Weiss, Analyse Harmonique Non-Commutative sur Certains Espaces. V, 160 pages. 1971. DM 16,–

Vol. 243: Japan-United States Seminar on Ordinary Differential and Functional Equations. Edited by M. Urabe. VIII, 332 pages. 1971. DM 26,–

Vol. 244: Séminaire Bourbaki – vol. 1970/71. Exposés 382–399. IV, 356 pages. 1971. DM 26,–

Vol. 245: D. E. Cohen, Groups of Cohomological Dimension One. V, 99 pages. 1972. DM 16,–

Vol. 246: Lectures on Rings and Modules. Tulane University Ring and Operator Theory Year, 1970–1971. Volume I. X, 661 pages. 1972. DM 40,–

Vol. 247: Lectures on Operator Algebras. Tulane University Ring and Operator Theory Year, 1970–1971. Volume II. XI, 786 pages. 1972. DM 40,–

Vol. 248: Lectures on the Applications of Sheaves to Ring Theory. Tulane University Ring and Operator Theory Year, 1970–1971. Volume III. VIII, 315 pages. 1971. DM 26,–

Vol. 249: Symposium on Algebraic Topology. Edited by P. J. Hilton. VII, 111 pages. 1971. DM 16,–

Vol. 250: B. Jónsson, Topics in Universal Algebra. VI, 220 pages. 1972. DM 20,–

Vol. 251: The Theory of Arithmetic Functions. Edited by A. A. Gioia and D. L. Goldsmith VI, 287 pages. 1972. DM 24,–

Vol. 252: D. A. Stone, Stratified Polyhedra. IX, 193 pages. 1972. DM 18,–

Vol. 253: V. Komkov, Optimal Control Theory for the Damping of Vibrations of Simple Elastic Systems. V, 240 pages. 1972. DM 20,–

Vol. 254: C. U. Jensen, Les Foncteurs Dérivés de $\varprojlim$ et leurs Applications en Théorie des Modules. V, 103 pages. 1972. DM 16,–

Vol. 255: Conference in Mathematical Logic – London '70. Edited by W. Hodges. VIII, 351 pages. 1972. DM 26,–

Vol. 256: C. A. Berenstein and M. A. Dostal, Analytically Uniform Spaces and their Applications to Convolution Equations. VII, 130 pages. 1972. DM 16,–

Vol. 257: R. B. Holmes, A Course on Optimization and Best Approximation. VIII, 233 pages. 1972. DM 20,–

Vol. 258: Séminaire de Probabilités VI. Edited by P. A. Meyer. VI, 253 pages. 1972. DM 22,–

Vol. 259: N. Moulis, Structures de Fredholm sur les Variétés Hilbertiennes. V, 123 pages. 1972. DM 16,–

Vol. 260: R. Godement and H. Jacquet, Zeta Functions of Simple Algebras. IX, 188 pages. 1972. DM 18,–

Vol. 261: A. Guichardet, Symmetric Hilbert Spaces and Related Topics. V, 197 pages. 1972. DM 18,–

Vol. 262: H. G. Zimmer, Computational Problems, Methods, and Results in Algebraic Number Theory. V, 103 pages. 1972. DM 16,–

Vol. 263: T. Parthasarathy, Selection Theorems and their Applications. VII, 101 pages. 1972. DM 16,–

Vol. 264: W. Messing, The Crystals Associated to Barsotti-Tate Groups: with Applications to Abelian Schemes. III, 190 pages. 1972. DM 18,–

Vol. 265: N. Saavedra Rivano, Catégories Tannakiennes. II, 418 pages. 1972. DM 26,–

Vol. 266: Conference on Harmonic Analysis. Edited by D. Gulick and R. L. Lipsman. VI, 323 pages. 1972. DM 24,–

Vol. 267: Numerische Lösung nichtlinearer partieller Differential- und Integro-Differentialgleichungen. Herausgegeben von R. Ansorge und W. Törnig, VI, 339 Seiten. 1972. DM 26,–

Vol. 268: C. G. Simader, On Dirichlet's Boundary Value Problem. IV, 238 pages. 1972. DM 20,–

Vol. 269: Théorie des Topos et Cohomologie Etale des Schémas. (SGA 4). Dirigé par M. Artin, A. Grothendieck et J. L. Verdier. XIX, 525 pages. 1972. DM 50,–

Vol. 270: Thèorie des Topos et Cohomologie Etle des Schémas. Tome 2. (SGA 4). Dirige par M. Artin, A. Grothendieck et J. L. Verdier. V, 418 pages. 1972. DM 50,–

Vol. 271: J. P. May, The Geometry of Iterated Loop Spaces. IX, 175 pages. 1972. DM 18,–

Vol. 272: K. R. Parthasarathy and K. Schmidt, Positive Definite Kernels, Continuous Tensor Products, and Central Limit Theorems of Probability Theory. VI, 107 pages. 1972. DM 16,–

Vol. 273: U. Seip, Kompakt erzeugte Vektorräume und Analysis. IX, 119 Seiten. 1972. DM 16,–

Vol. 274: Toposes, Algebraic Geometry and Logic. Edited by. F. W. Lawvere. VI, 189 pages. 1972. DM 18,–

Vol. 275: Séminaire Pierre Lelong (Analyse) Année 1970–1971. VI, 181 pages. 1972. DM 18,–

Vol. 276: A. Borel, Représentations de Groupes Localement Compacts. V, 98 pages. 1972. DM 16,–

Vol. 277: Séminaire Banach. Edité par C. Houzel. VII, 229 pages. 1972. DM 20,–